Design an RP2040 board
with KiCad, 1st Edition

Design an RP2040 board with KiCad
by Jo Hinchliffe and Ben Everard
ISBN: 978-1-916868-13-7
Copyright © 2024 Jo Hinchliffe and Ben Everard
Printed in the United Kingdom
Published by Raspberry Pi Ltd., 194 Science Park, Cambridge, CB4 0AB

Editors: Ben Everard, Brian Jepson
Interior Designer: Sara Parodi
Production: Nellie McKesson
Photographer: Brian O'Halloran
Illustrator: Sam Alder
Graphics Editor: Natalie Turner
Publishing Director: Brian Jepson
Head of Design: Jack Willis
CEO: Eben Upton

September 2024: First Edition

Table of Contents

Welcome

KiCad is an amazing piece of free and open-source software that allows anyone, with some time and effort, to make high-quality PCB designs. Couple this amazing software with numerous PCB fabrication companies and even PCBA services — companies that will make and assemble your PCB designs — and there's never been a better time to get into this aspect of making.

This book provides a gentle introduction to PCB design using the RP2040 microcontroller chip (the same chip that's at the heart of Raspberry Pi Pico). You'll learn the basics of creating schematics and PCB designs in KiCad and learn how to work with artifacts such as component footprints that you create yourself (or get from another source). You'll find out how to get a PCB design manufactured — and populated with surface-mount components.

You'll also learn how to make your PCBs stand out from more generic boards, whether by adding your own artwork, using the PCB as a structural component, or augmenting your design with 3D-printed parts. You'll also find out about the difference PCB materials available, including flexible PCBs, and learn tips and tricks for working with fabricators to make sure your boards come out as intended.

You can find this book's example code, errata, and other resources in its GitHub repository at **hsmag.cc/kicad_book_files**. If you've found what you believe is a mistake or error in the book, please let us know by opening an issue in that GitHub repository.

About the authors

Jo Hinchliffe (AKA Concretedog) is a constant tinkerer and is passionate about all things DIY space. He loves designing and scratch-building both model and high-power rockets and releases the designs and components as open source. He also has a shed full of lathes and milling machines and CNC kit!

Ben Everard enjoys working at the intersection of art and technology. He has a particular interest in light and his primary reason for making PCBs is to find more ways of adding LEDs to things. He lives in a house in Bristol with his wife, two daughters and too many animals.

Colophon

Raspberry Pi is an affordable way to do something useful, or to do something fun.

Democratising technology — providing access to tools — has been our motivation since the Raspberry Pi project began. By driving down the cost of general-purpose computing to below $5, we've opened up the ability for anybody to use computers in projects that used to require prohibitive amounts of capital. Today, with barriers to entry being removed, we see Raspberry Pi computers being used everywhere from interactive museum exhibits and schools to national postal sorting offices and government call centres. Kitchen table businesses all over the world have been able to scale and find success in a way that just wasn't possible in a world where integrating technology meant spending large sums on laptops and PCs.

Raspberry Pi removes the high entry cost to computing for people across all demographics: while children can benefit from a computing education that previously wasn't open to them, many adults have also historically been priced out of using computers for enterprise, entertainment, and creativity.

Raspberry Pi eliminates those barriers.

Raspberry Pi Press

store.rpipress.cc

Raspberry Pi Press is your essential bookshelf for computing, gaming, and hands-on making. We are the publishing imprint of Raspberry Pi Ltd. From building a PC to building a cabinet, discover your passion, learn new skills, and make awesome stuff with our extensive range of books and magazines.

The MagPi

magpi.raspberrypi.com

The MagPi is the official Raspberry Pi magazine. Written for the Raspberry Pi community, it is packed with Pi-themed projects, computing and electronics tutorials, how-to guides, and the latest community news and events.

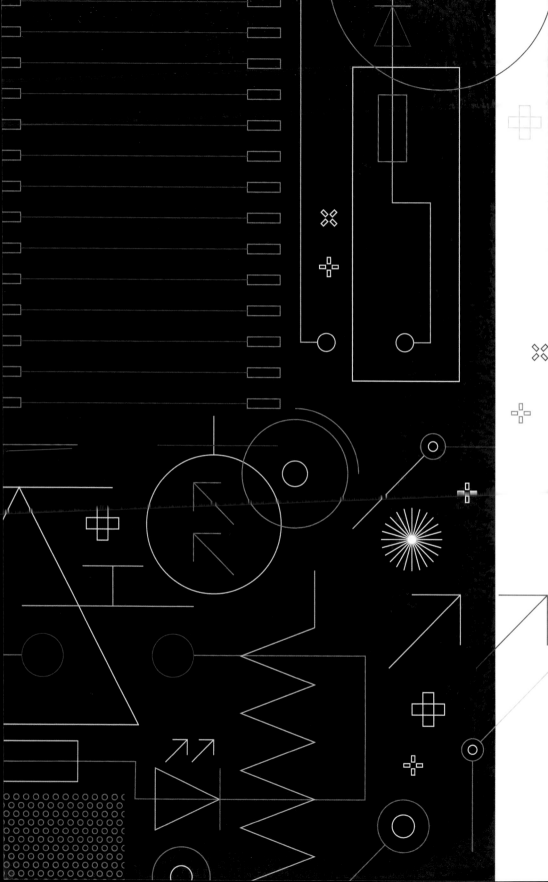

Chapter 1

Working with schematics

Before you design the PCB, you must design the circuit

PCB design has a steep learning curve. With that in mind, we want to start this book with a hack. In the first two chapters, you'll create a PCB design to the point of getting it manufactured. To ease up on the learning curve, you will cut some corners. Don't worry, we'll explain what corner were cut, and you'll learn the correct approach in subsequent chapters, once you have the basics down.

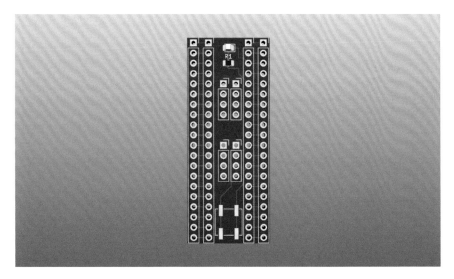

Figure 1-1   The board you'll start to design in this chapter

Start by downloading and installing KiCad from **kicad.org** — at the time of writing, the stable version was 8. KiCad is available across a wide range of operating systems – Windows, macOS, and lots of Linux distributions.

With it installed, click on the main blue **Ki** KiCad icon to open the main application. You should see a screen that looks like **Figure 1-2**. KiCad isn't really a single application, it's more like a suite of applications that work together. While you can jump into any application from here, the most common workflow for KiCad is to first work in the Schematic Editor and after creating a schematic, move to the PCB Editor to lay out the PCB physically.

You'll create an add-on board for the Raspberry Pi Pico. It's going to be a proto-typing or 'kludge' board with not too many features on it. It's going to have all the Pico pins broken out to through-hole pads, it will have a power-indicating LED, and it will have a reset button. After this, any spare area on the board will have a grid of simple through-hole pads (sometimes called a prototyping area) on it to allow you to connect experiments to the board. It's a bit rudi-mentary but will help you learn some KiCad basics.

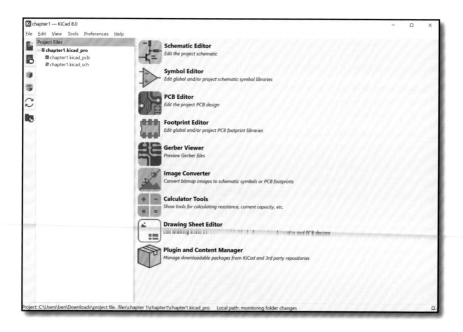

Figure 1-2   The opening page of KiCad with the different applications that make up the KiCad suite listed

Your first action should be to set up a new project. Click **File > New Project** to create a new project. It's worth putting projects into their own folder as each KiCad project generates around five project files and a folder in which it au-tomatically generates some project backup files.

Once the new project is created, open the Schematic Editor application by clicking the top icon. The first time you run the Schematic Editor, KiCad will ask you to configure your global symbol library table. Choose **Copy default global symbol library table (recommended)** and click OK.

Quick Tip

If you hover over any tool icon in KiCad, you get a description of the tool. We'll use these tooltips to describe tool icons throughout this book.

You'll see a blank page ready for you to draw a schematic (**Figure 1-3**). In the lower right-hand corner, you will see a small collection of text boxes which include various fields for the name of the sheet, the revision number of the schematic, and more. If you click somewhere on this section and then press the E key, the **Page Settings** dialogue will appear. You can edit this to add any text details or comments you want to add to this section, but you can also change the page dimensions (it defaults to A4), the orientation, and more. Experiment inserting text into the **Page Settings** window to see where the text appears on the schematic.

Figure 1-3 A blank Schematic sheet with text boxes for various schematic labelling and version numbering

With your page set up, you can now begin to add schematic symbols to the schematic. The most correct way to make a Pico add-on board would be to place a Pico in the schematic and connect everything to it, but you are going to use a workaround to do this in a much simpler way. The main idea of the target add-on board is for you to be able to solder in rows of header sockets so that you can connect it onto the pins of a Pico. In turn, you also still want to be able to solder to those pins, so you need each side row of pins to be broken out to another collection of pads.

To do this add two 20-pin connectors, one on each side of the schematic. To add the first one, click the **Add a Symbol** icon — the third icon down on the right-hand column. The first time you click this icon, it may take a few seconds for the libraries of schematic symbols to load and you will be offered a choice of which library to load — select the option **Copy default global footprint library table (recommended)**.

Once loaded, the **Choose Symbol** dialogue appears. On the left-hand side, you'll see a list of items, each of which is a *library* (a group of schematic symbols). These are grouped as related items. So, you might find, at the top of the list, **4xxx**. If you click on the drop-down menu button next to it, it will reveal a library full of the CMOS 4000 series of logic chips. You can manually scroll through the libraries and look through the items they contain, but you can also search the libraries to find what you need.

You are going to add a connector symbol that represents the 20-pin header you eventually want to be able to fit. To do this, type **conn** into the search bar and, as you type, you should notice the list items are now all the items that start **conn**. Scroll down the list and select the item that reads **Conn_01x20_Pin**, and then click the OK button in the lower right-hand corner of the window (**Figure 1-4**).

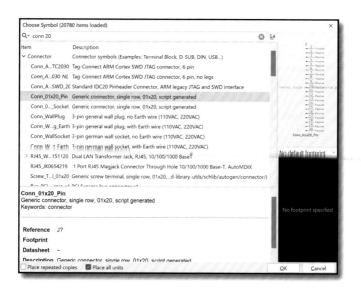

Figure 1-4   The Choose Symbol dialogue that allows you to search for schematic symbols

The **Choose Symbol** window will close, and you will be back in the Schematic Editor, where you should have a symbol for the 20-pin socket attached to your pointer. Move the pointer to where you want to place the connector and click to place it.

A sackload of sockets

You are going to place three more of 20 pin connectors into the schematic. Click the **Add a Symbol** tool icon again, but notice that, as the library list area populates, you now have a **Recently Used** area at the top with the 20-pin connector listed underneath (you may need to scroll the library list area to the top to see this). Click this listing and then **OK**, then place the next connector in the schematic. Repeat this until you have placed four of the connectors.

Next, use the **M** (Move) hotkey to move the connectors so that there are two pairs of connectors next to each other. One pair of headers corresponds to the left side of the Pico with one header being the holes for the Pico's pins and one header being holes to attach additional hardware. The other pair is the same thing but on the right-hand side.

Keyboard warrior

Whilst you can do everything in KiCad with a mouse and pointer, it's worth learning a few quick keyboard shortcuts to make life easier. Many keystrokes offer the same functions in both the **Schematic Editor** and the **PCB Editor**.

The first useful ones are the **F1** and **F2** keys for zooming. The **F1** key, when pressed, will zoom in, and **F2** will zoom out. Note that both are centred around where your pointer is. With a little practice, it becomes second nature to move the pointer and zoom in and out to get the view you need.

Next up are the **M** and **R** keys. If you select an object in the Schematic Editor or the PCB Editor, you can press **M** and then that object will move until you click to place the object again. In the PCB Editor, you can select smaller parts than the complete footprint of a component. You can, for example, select and move silkscreen labels and more. If you want to select the whole component, selecting a component pad will usually select the entire footprint.

With the **R** key, you can rotate a selected item in either editor, again single-clicking to place the part when rotated correctly. Note that you can rotate the item repeatedly to get to the orientation you require.

Finally, another useful shortcut is the **E** key. With an item selected, pressing the **E** key allows you to edit that item's properties. This can be any number of editable parameters, from labels to pad sizes to hole dimensions and more.

When using hot keys like **M**, place the cursor over the symbol and then press the key — there's no need to highlight the symbol first. The hot key can work on the complete symbol as well as individual parts of it. The easiest way of grabbing the whole connector is to place your cursor in a bit of blank space inside the component, then press the key. It might take a few attempts to get the hang of it. If you find it hard to get your cursor in the right place, try zooming in a little.

You should now have two pairs of connectors, but to make it easier to connect the wires, you'll want the little circles on the inside of the pair — these are the bits you'll connect the wires to and it's a bit tidier if the wires don't go across the symbol. You can do this using the **X** (Mirror Horizontally) hot key. You'll only need to do this for the rightmost connector in each pair.

Although you don't need it this time, the **R** (Rotate) hotkey is often also useful for getting your parts in a useful place.

Using the **Add a Wire** (click on the icon in the toolbox or press **W**) tool, you can click on one of the connector pin circles and then draw a connector wire line across to the opposite connector pin (**Figure 1-5**). If you make a mistake, you can use **CTRL+Z** to undo the last action — or, to cancel the wire mid-draw, you can right-click and select **Cancel** from the drop-down menu.

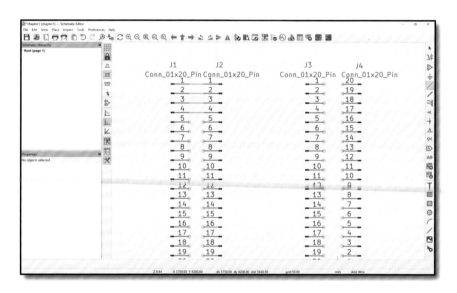

Figure 1-5   Adding wire connections between the pins of the schematic symbols

Continue to wire between each of the opposite pins on each connector pairs until they are all connected together. Notice that each of the connector symbols have a couple of text references as part of them. It can be good practice to edit these so that they are useful and help you keep track of what things are. Select the whole of one of the connector symbols (use the selection tool to draw a box over the entire symbol), put the cursor over the symbol, and then press **E**.

You should now see a **Symbol Properties** dialogue box (**Figure 1-6**). You can now edit the **Reference** (if required) and the **Value** text entries. You should give each connector a descriptive **Value** such as **Left Pico**. This is useful when you lay out the PCB in the PCB Editor, as it ensures you can identify the various connectors correctly.

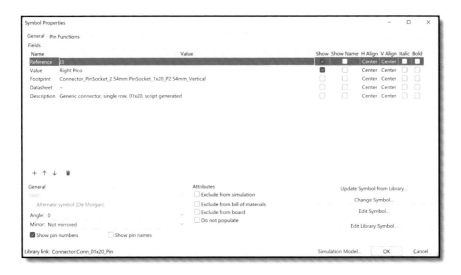

Figure 1-6   Editing a symbol's properties allows you to give symbols more meaningful names

To finish the schematic, add a resistor and an LED and connect them to pins 38 (**GND**) and 36 (**3V3 (OUT)**) on Pico. These pins are the third pin and fifth pin down on the right-hand connector. Use the same techniques you used earlier to choose a symbol, searching **R** for a resistor and **LED** for an LED symbol. Add wires to connect the circuit, as shown in **Figure 1-7**. Use the **R** hotkey to rotate symbols if needed.

Switched on

Next, add a switch symbol which you will wire in as a reset button for the Pico.

Adding a reset button is a good example of a way that KiCad works differently to some other Electronic Design Automation (EDA) tools. In some EDA tools, the symbol you choose at the schematic level defines exactly the hardware package of the electronic component that will be on the PCB.

In other EDA tools, you wouldn't place a generic resistor symbol, you would place a resistor symbol linked to a specific resistor, say a 6mm long 1/4-watt carbon resistor placed horizontally. This would mean that, if you wanted to change the component on the PCB, you have to change the schematic. In KiCad,

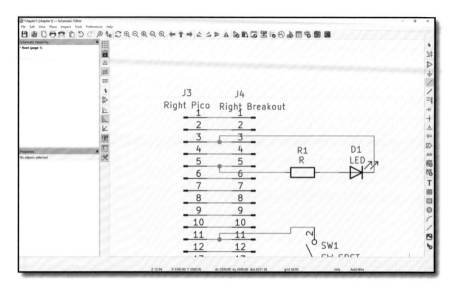

Figure 1-7 Adding a resistor and LED to the schematic

the schematic symbol is only assigned a *footprint* — this is why you are going to place a symbol for a single-pole single-throw switch (SPST), but the actual component can be any SPST switch or any button package. In this case, you'll choose some type of momentary push-button for the reset. This way of working means that if you find you need to replace a component with a different type (for example, if you can't find a component in stock), you simply change the footprint associated with the symbol and don't have to edit your schematic. It also makes the schematics concise and readable, which is important if you want to share your design.

To add the switch, search the **Choose a Symbol** dialogue with **sw_spst** to take you directly to the single-pole single-throw switch and again connect wires, as shown in **Figure 1-8**. Finally, add four connectors to act as prototyping areas on the board; search for **conn_01x04** to take you directly to this symbol. You should now have a schematic that looks like **Figure 1-9**.

Let's set up the associated footprints for each schematic symbol. Go to **Tools > Assign Footprints**. The **Assign Footprints** dialogue (**Figure 1-10**) appears with a list of footprint libraries down the left-hand side, a centre section with a list of schematic symbols in the project, and a right-hand column of filtered footprint results. There are three icons at the top of the window called **Footprint Filters** and, as you'll see in a moment, it's helpful to select the second of these (**Filter by pin count**). As you get used to component filtering, you might find you prefer to use different combinations of these filtering tools.

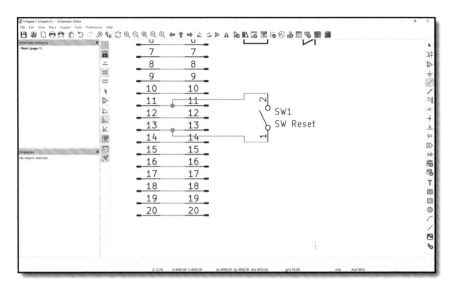

Figure 1-8 Adding a single-pole single-throw switch to the schematic to work as a reset button

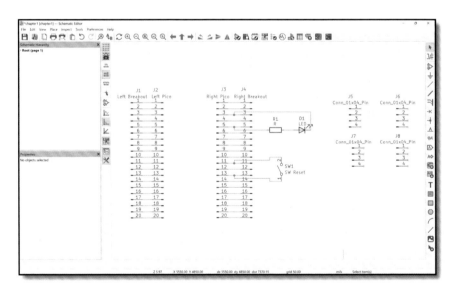

Figure 1-9 The complete circuit schematic

Highlight one of the 20-pin connector components in the centre section. You need to find a footprint that has the same number of pins and has a footprint of a through-hole pad for each pin. As the Raspberry Pi Pico pins are

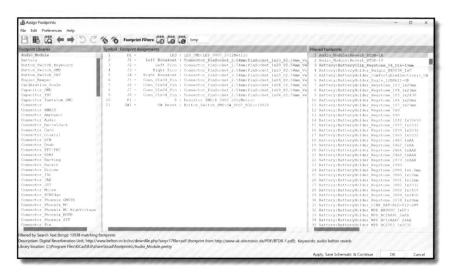

Figure 1-10   The Assign Footprints dialogue

spaced at the common 2.54mm pitch between each pin, you also need to make sure the footprint is spaced similarly. You should have a filtered list on the right-hand side — it should be filtered to only contain items that would attach to the 20-pin connector-symbol. Choose **Connector_PinSocket_2.54mm:Pinsocket_1x20_P2.54mm_vertical** from the connector footprint library. Click once on the item to highlight it in the list, then use the **View the selected footprint in the footprint viewer** tool icon (third icon from the left) to open a window showing the PCB footprint layout design to check if it looks correct. Double-click the highlighted footprint to associate it with the selected schematic symbol in the central list. Continue and add the same footprint to all four connector symbol listings (we'll discuss later why that isn't totally correct, but it's fine for this starter project). Click the **Apply, save schematic and continue button**.

You'll also need to add surface-mount components for the LED, resistor, and the button; you'll choose SMD package sizes and footprints that are easy to hand-solder. For the LED, choose the **LED_SMD:LED_0805_2012Metric** footprint; for the resistor, the **Resistor_SMD:R_0805_2012Metric;** and for the switch, a **Button_Switch_SMD:SW_SPST_B3SL-1002P** component. For the detached 4-pin connectors, select **Connector_PinSocket_2.54mm:PinSocket_1x04_P2.54mm_Vertical** for each connector. With all those footprints assigned, click **Apply, save schematic and continue**, then click OK.

In the next chapter, you will import the list of component footprints into the PCB Editor and physically lay out the board design to get it ready for fabrication.

Adding parts

As you can tell from this chapter, KiCad uses schematic symbols and component footprints to create schematics and PCB designs. You should have a lot of standard libraries built in, and everything in this project is just using these libraries. Of course though, you can create your own schematic symbols, component footprints, and even 3D models of components and incorporate these into your own designs creating custom libraries. We'll cover this, and lots more in future chapters.

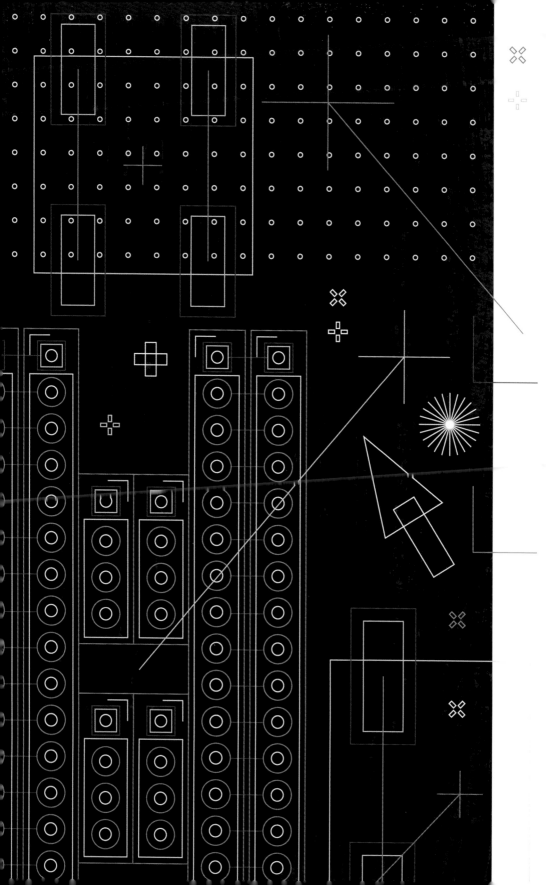

Chapter 2

Laying out a PCB

Turn a schematic into a physical design

In the last chapter, you created a schematic for a Raspberry Pi Pico add-on board, and assigned each schematic symbol a footprint which represents the individual component using. In this chapter, you'll finish this starter project by laying out the PCB design, ready for manufacture.

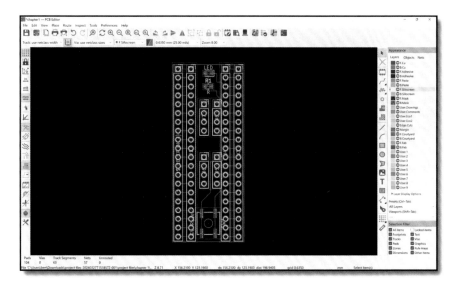

Figure 2-1   The routed PCB you'll design in this chapter

Open the previous project in KiCad and select the **PCB Editor** icon from the available applications. If you have opened the project in the Schematic Editor, jump to the PCB Editor by clicking its icon in the Schematic Editor toolbar.

Having opened the PCB Editor, you should see a similar page to the Schematic Editor, but in black (**Figure 2-2**). You'll notice it is blank, so the first action is to pull in the footprints.

Let's change one setting before placing the footprints. The PCB editor has a grid feature which snaps footprints at useful points. In the current case, the pin socket footprints will snap with the centres of the pad holes on the grid. As Pico conforms to 2.54mm pin spacings, it's useful at this point to set the board editor grid to **2.54mm (100 mils)**. You can select this from the centre drop-down menu on the upper toolbar.

Next, click the **Update PCB from Schematic** icon (4th icon from the right in the top toolbar or press **F8**). This opens a dialogue, and you can simply click the **Update PCB** button and then click the **Close** button (**Figure 2-3**). The **Update PCB from Schematic** dialogue brings in the component footprints and connectivity and can be used to apply future changes to a schematic design. Once back in the PCB Editor, you now should have the design as a collection of footprints attached to your mouse pointer. Click to place the component bundle somewhere in the middle of the page (**Figure 2-4**).

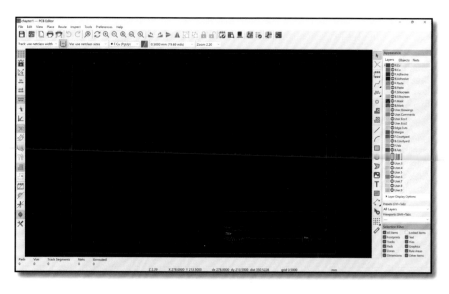

Figure 2-2 The blank PCB Editor ready to import the footprints

Mil is a unit of measurement that's equal to 0.001 inches. It's common in the PCB world, and easily confused with mm.

You can now align the two pairs of pin socket footprints so that they are in line with each other horizontally.

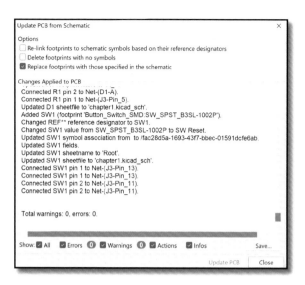

Figure 2-3   The **Update PCB from Schematic** dialogue

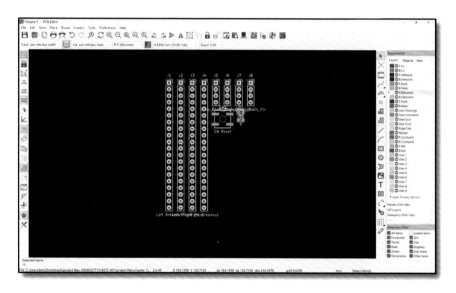

Figure 2-4   The component footprints imported, with the rat's nest a representation of the connections between components and pads made of small lines

Start by moving the footprint labelled Left Pico: make sure the **Select item(s)** tool is selected on the right-hand toolbar, then click on the centre of a pad on the footprint and press **M** (similar to how you worked with the Schematic Editor). Move it about ten grid dots and click to place it. As you move it, you should notice a collection of small white lines connecting the pads between the two connected footprints — this is called *the rat's nest*. You will remove these white lines when you lay the trace connections.

It's important that the board matches up to Pico, so we used the dimensional diagram (**Figure 2-5**) taken from the Pico Data Sheet, available at **hsmag.cc/PicoDatasheet**. You'll notice that the distance between the vertical rows of pins is 17.78mm — this equates to 7 × 2.54mm. This spacing is common in electronics — it's why your Pico fits perfectly into a prototyping breadboard.

Click the Right Pico footprint, and again press **M**. Move the mouse to align the Right Pico footprint on top of the Left Pico footprint, and then move the Right Pico footprint seven grid spots to the right.

For the remainder of the component placing, you can get things to fit a bit better if you change the grid size to **1.2700mm (50 mils)**. You can now arrange the Left Breakout footprint to be two grid spots to the right of the Left Pico, and the Right Breakout footprint two grid spots to the left of the Right Pico. The outside pair of footprints should now match Pico's pin positions, and the inner footprints should be close to them, and parallel. You can use the Measure tool (the ruler-shaped icon at the bottom of the right-hand toolbar) to confirm that the distance between the outer two footprints is 17.78mm.

So far, you should have only been working on the **F.Cu (PgUp)** layer, which is the top copper layer on the PCB board. Before you route the tracks between the footprints, double-check that you are on this layer by checking the drop-down menu on the top toolbar.

Leave only footprints

Next, you can wire the pads on the opposite footprints together (**Figure 2-6**). To do this, select the **Route Tracks (X)** icon, then click the centre of a pad and drag the track over to the centre of the opposite pad. The track should be red, indicating that it's on the top copper layer. If you, for example, were to move to the **B.Cu (PgDn)** layer, the default colour for tracks on the lower copper

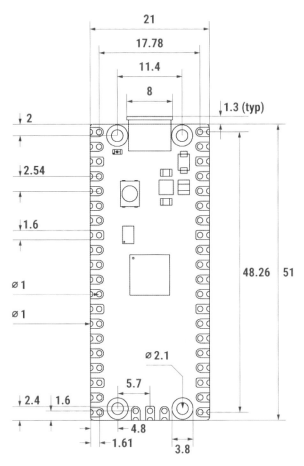

Figure 2-5 A technical drawing of the Pico taken from the data sheet gives you the dimensional information you need

layer is blue. Continue and connect all pads together, noticing that the rat's nest lines disappear as you do so.

Referring to the Pico technical drawing, it's a good time to define the shape and edges of the board. To do this, you can use the uppermost drop-down menu to move from the **F.Cu (PgUp)** layer to the **Edge.Cuts** layer. On this layer, you will use the **Draw a Rectangle** tool to create a cutout shape for the board. But before you select this tool, switch the grid to a more useful spacing. Select the 1mm grid spacing, then click the rectangle tool. Click anywhere in the PCB Editor page and drag a rectangle (**Figure 2-7**). It doesn't have to be over the other components because you'll move it in the next step.

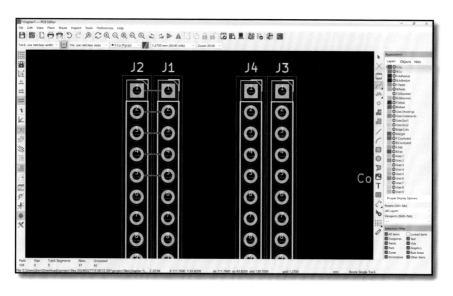

Figure 2-6   Connecting up the pads on the board

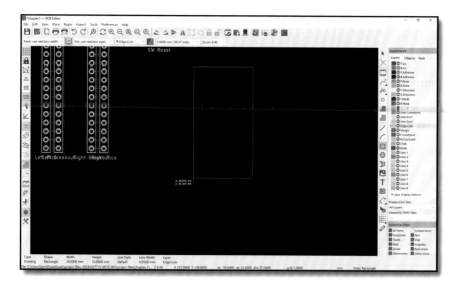

Figure 2-7   The outline of the board is a rectangle in the Edge.Cuts layer

The rectangle should snap to the grid, and you should drag it out until the labels tell you that it's 21mm wide by 51mm high. Click one more time to create the rectangle. Moving back to the **Select Items (s)** tool, you should now select the rectangle and press the **M** key to move it into position. Notice from the Pico technical drawing that the outside edge of the Pico sits 1.61mm from the centre of the pin pad position. To position this accurately, you should reduce the grid spacing to the smallest listed and use the following technique to measure distances on the page.

If you press the space bar at any time when in the PCB Editor, you might notice that values labelled **dx, dy**, and **dist** are set to zero. This is very useful as you can place your pointer at a point, say the centre of the top rightmost pad, press the **SPACE** bar to create a zero or datum point, and then move the pointer to, in this case, the edge cut rectangle. You can then use the **dx** and **dy** values to help you position this, or anything you need to, accurately.

Quick Tip

You can actually look at the board in the 3D Viewer before adding an edge cut, and it will render to a rectangle size that is the smallest that can accommodate all the footprints currently in the PCB.

Once you have positioned the board edge rectangle correctly, you can get a first glimpse of the board in the 3D Viewer. To view the PCB, click **View > 3D Viewer**. You'll see the board, but you'll notice that all the 3D models of the header sockets are all placed into the rows of holes. In reality, you would only want headers on the outer rows installed on the back of the board, with the inner rows left unpopulated, or possibly populated with header pins. You could edit the board and the component footprints to reflect this, but for this simple example, you can click the **Toggle 3D Models for Through Hole type Components (T)** icon to turn off the models (**Figure 2-8**). In future chapters, we'll explore not only using correct 3D models but will look at how you can add custom 3D models too.

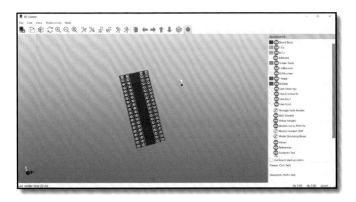

Figure 2-8 A first glimpse of the rendered PCB board

Keep your distance

You can now continue to arrange and wire the remaining components. In order to fit the components, you'll need to switch to the 0.635mm grid, and you may need to rotate the switch using the **R** hotkey. You'll find that you need to overlap some of the footprints slightly. When doing this, you need to understand what each part of the footprint is showing.

The purple box around it shows the maximum size of the component. In normal usage, these shouldn't overlap. However, in the case of the headers, you're not actually going to add physical headers (you're just using this footprint to give you a series of through holes in the right place). Therefore, it's fine to overlap these purple rectangles slightly. The pale yellow lines are silkscreen — basically, graphics that are printed onto the PCB. Again, for this board, these can overlap slightly provided it won't interfere with actual components. As you move onto boards with more actual parts on them, you'll need to be much more careful about things not getting in the way of each other.

Assigned and sealed

One point mentioned earlier in this project is the idea that KiCad schematic symbols are generic — you assign the hardware footprint to them using the **Assign Footprint** tools. Here's the advantage of this. Say you have a PCB layout completed and ready for manufacture, but after checking, you realise that the **B3SL-1002P** button package is not available or in stock anywhere. You can simply go back to the Schematic Editor and click the **Assign Footprint** tool icon. You can then edit the **SW_SPST** symbol to have a different, and hopefully, in stock, component. For example, select the **B3SL-1022P** footprint package, double-clicking it to ensure it is added to the centre console list. You can then click **Apply ▸ Save Schematic ▸ Continue**, and then click **OK** to close the dialogue. Moving back to the PCB Editor, you can once again click the **Update PCB with changes made to the schematic (F8)** icon, or press **F8** on the keyboard and the board will update with the replacement footprint added. If, as in this example, you replace the part with another part that is virtually the same footprint, you might not need to rewire the traces, but, of course, you should check and adjust connections and positioning as required.

For the LED and the button, you need to wire traces directly to the pin connections to the Pico. You'll find that you must weave the traces through other components. Each trace has to avoid pads and through holes of other components but can travel through the footprint as long as it misses these. A single right-click adds a bend point to a wire.

You might find that you need to rotate a component through 180 degrees to make the connections come out of the right sides. You might also find that it works better if you decrease the grid size to allow you to place the tracks more precisely.

You have quite a crowded board, and there's not much space for text. The labels that you used to remind yourself which header is which are no longer needed once you've laid out the PCB, so you can simple delete these (click in the text and press the **DELETE** key).

Finally, how do you get this board made? Well, there are lots of options — we will look at many of them throughout this book. There are many services available from companies all over the world to manufacture and even assemble your boards.

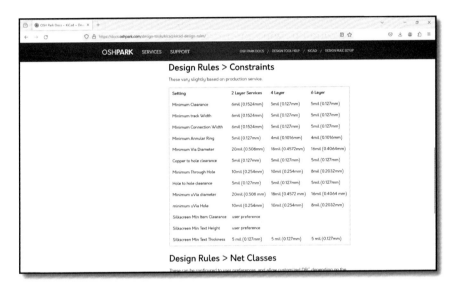

Figure 2-9 We use numerous PCB fabricators in this book, but a great place to start, where you can directly upload KiCad PCB files, is OSHPark

These services may need different approaches in terms of what information and files you need to upload to get the job done. Often you'll need to export Gerber files for each layer of the PCB, and also export *drill files* which show the position and size of holes. Some companies might have limitations on what size holes they can produce and what tolerance they can produce the board too. We'll explore this further in future chapters, but if you want to get this board fabricated to an excellent standard, upload the file that ends with .**kicad_pcb** to the OSHPark website (**oshpark.com**). The website and service are brilliant. In-browser, it creates numerous renders of your board, which you can then inspect and check to see if they are correct before adding the PCB to your cart to be manufactured. In a few weeks — often less — you'll have 'Perfect Purple PCBs' through your door.

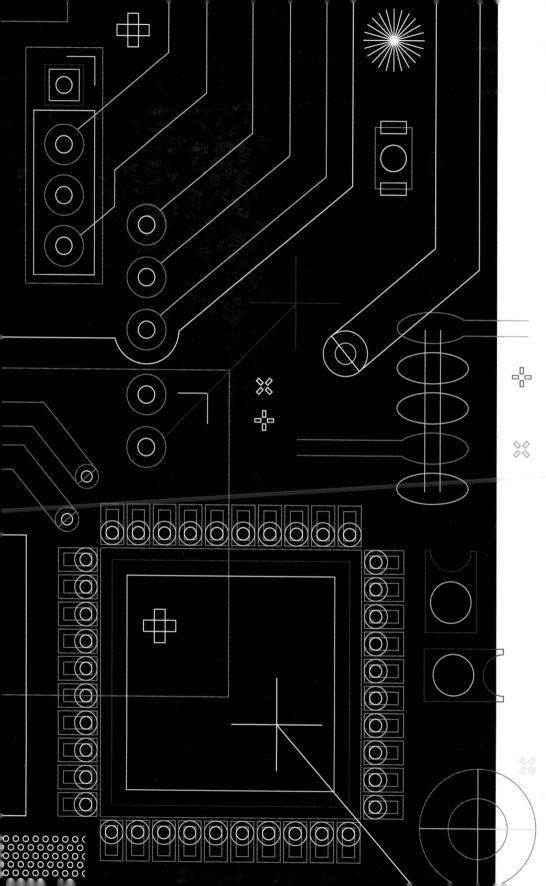

Chapter 3

Libraries, symbols and footprints

Manage your components

In the first two chapters of this book, we covered enough to make a basic PCB suitable for simple circuits — a breakout board with some added components. In this chapter, we show how you can expand the capabilities of what you can make in KiCad. We'll explore both creating libraries and symbols from scratch, but also importing and using component footprints and schematic symbols from other sources. You'll also improve the quality of the boards by using *flooded* areas for common connections, such as all the circuit points that are connected to ground.

Figure 3-1 The completed board, ready to receive a Solder Party Stamp and a Pimoroni BMP280 module

This chapter uses a relatively simple design to show these techniques — making a PCB that essentially has two modules on board. Quite often, when building a project, you'll work with electronics modules on breadboards. This helps test out things and get the wiring correct, but it's not particularly suitable for long-term use, because wires can fall out and parts go astray. A custom PCB can be the perfect way to take a breadboard project to a permanent home that's more rugged and usable. While this chapter won't cover every step in complete detail, the project files are available at **hsmag.cc/kicad_book_files**, and the knowledge and techniques you learned in the first two chapters, in combination with this section, should let you recreate this project.

You should download this project and keep it as a reference, but start with a blank project as you build up to it.

This project connects a Solder Party Stamp and a Pimoroni BMP280 module so that the resulting board can be used to measure and log temperature and barometric pressure. The Solder Party Stamp is an excellent board that has an RP2040 at its heart and is operationally similar to a Raspberry Pi Pico. The RP2040 is fully broken out to header pins which are *castellated*, so you can also solder the Stamp onto a PCB's pads without having to use header pins. The Stamp has the USB connection broken out as well as on-board LiPo charging. This means if you add a USB connection, you can also add a LiPo cell and make the project stand-alone. The Solder Party Stamp is well-documented and is open source. Solder Party has also published KiCad schematic symbols and a PCB footprint for the Stamp, therefore you can use it to learn how to add libraries and import these useful items into KiCad.

To begin, go to the following link where you will find the Solder Party Stamp library components: **hsmag.cc/StampFootprints**. Click the drop-down menu on the green **Code** button, then select the **Download ZIP** option to download the libraries. Unzip the files somewhere on your machine. In the collection of folders you just unzipped, move the entire **KiCad** folder (N.B. not the **KiCad 5** folder) to wherever you want to store your additional external KiCad libraries. You should have a folder set up in your home directory for this.

Leveraging libraries

Open KiCad and, in the main page, click the **Preferences** drop-down menu and then select the **Manage Symbol Libraries** option. This should open a window with two tabs: the **Global Libraries** tab and the **Project Specific** tab, (**Figure 3-2**). Ensuring you are on the **Global Libraries** tab, find and click the small folder icon. Navigate to the folder you downloaded and open it to find a folder called **KiCad_stamp_lib**. Open this folder and select **RP2040_Stamp.kicad.sym** and then click **Open**.

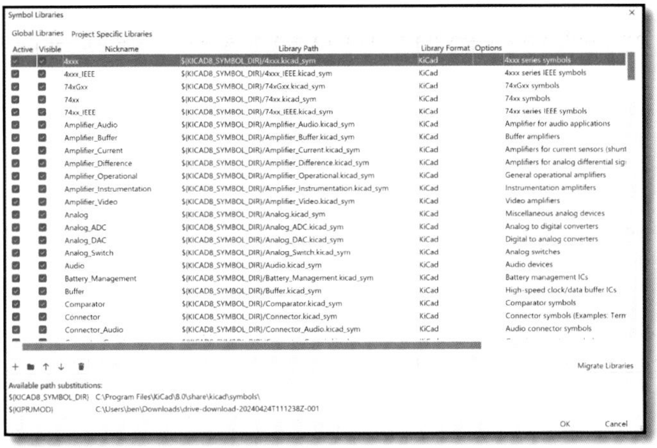

Symbol Libraries

Global Libraries   Project Specific Libraries

Active	Visible	Nickname	Library Path	Library Format	Options
		4xxx	$(KICAD8_SYMBOL_DIR)/4xxx.kicad_sym	KiCad	4xxx series symbols
		4xxx_IEEE	$(KICAD8_SYMBOL_DIR)/4xxx_IEEE.kicad_sym	KiCad	4xxx series IEEE symbols
		74xGxx	$(KICAD8_SYMBOL_DIR)/74xGxx.kicad_sym	KiCad	74xGxx symbols
		74xx	$(KICAD8_SYMBOL_DIR)/74xx.kicad_sym	KiCad	74xx symbols
		74xx_IEEE	$(KICAD8_SYMBOL_DIR)/74xx_IEEE.kicad_sym	KiCad	74xx series IEEE symbols
		Amplifier_Audio	$(KICAD8_SYMBOL_DIR)/Amplifier_Audio.kicad_sym	KiCad	Amplifier for audio applications
		Amplifier_Buffer	$(KICAD8_SYMBOL_DIR)/Amplifier_Buffer.kicad_sym	KiCad	Buffer amplifiers
		Amplifier_Current	$(KICAD8_SYMBOL_DIR)/Amplifier_Current.kicad_sym	KiCad	Amplifiers for current sensors (shunt
		Amplifier_Difference	$(KICAD8_SYMBOL_DIR)/Amplifier_Difference.kicad_sym	KiCad	Amplifiers for analog differential sig
		Amplifier_Operational	$(KICAD8_SYMBOL_DIR)/Amplifier_Operational.kicad_sym	KiCad	General operational amplifiers
		Amplifier_Instrumentation	$(KICAD8_SYMBOL_DIR)/Amplifier_Instrumentation.kicad_sym	KiCad	Instrumentation amplifiers
		Amplifier_Video	$(KICAD8_SYMBOL_DIR)/Amplifier_Video.kicad_sym	KiCad	Video amplifiers
		Analog	$(KICAD8_SYMBOL_DIR)/Analog.kicad_sym	KiCad	Miscellaneous analog devices
		Analog_ADC	$(KICAD8_SYMBOL_DIR)/Analog_ADC.kicad_sym	KiCad	Analog to digital converters
		Analog_DAC	$(KICAD8_SYMBOL_DIR)/Analog_DAC.kicad_sym	KiCad	Digital to analog converters
		Analog_Switch	$(KICAD8_SYMBOL_DIR)/Analog_Switch.kicad_sym	KiCad	Analog switches
		Audio	$(KICAD8_SYMBOL_DIR)/Audio.kicad_sym	KiCad	Audio devices
		Battery_Management	$(KICAD8_SYMBOL_DIR)/Battery_Management.kicad_sym	KiCad	Battery management ICs
		Buffer	$(KICAD8_SYMBOL_DIR)/Buffer.kicad_sym	KiCad	High-speed clock/data buffer ICs
		Comparator	$(KICAD8_SYMBOL_DIR)/Comparator.kicad_sym	KiCad	Comparator symbols
		Connector	$(KICAD8_SYMBOL_DIR)/Connector.kicad_sym	KiCad	Connector symbols (Examples: Term
		Connector_Audio	$(KICAD8_SYMBOL_DIR)/Connector_Audio.kicad_sym	KiCad	Audio connector symbols

Migrate Libraries

Available path substitutions:
$(KICAD8_SYMBOL_DIR)   C:\Program Files\KiCad\8.0\share\kicad\symbols\
$(KIPRJMOD)   C:\Users\ben\Downloads\drive-download-20240424T111238Z-001

OK   Cancel

Figure 3-2   The **Symbol Libraries** dialogue where you can add or remove schematic symbol libraries

You should now see a new library listed at the bottom of the **Global Libraries** tab called **RP2040_Stamp**. If you create a new project and open the Schematic Editor, you can now use the **Add a Symbol** tool to place a Solder Party Stamp symbol into the schematic. You can do this by searching for **RP2040** and making sure you select the symbol from the **RP2040_Stamp library** (rather than the stock **RP2040** symbol) or by scrolling down the list of libraries, selecting **RP2040_Stamp**, and then selecting the **RP2040_Stamp** symbol.

It's a similar experience to add a footprint library. Again, in the KiCad landing page, click **Preferences** and then **Manage Footprint Libraries**. Again, on the **Global Libraries** tab, click the small folder icon (**Figure 3-3**). Navigate to the folder you downloaded and find **KiCad_stamp_lib** — open that folder once more but, this time, select the **RP2040_Stamp.pretty** folder and click Open. You should see three files inside, but you don't need to select any particular one — just click **Open** again. Now, back on the **Global Libraries** tab, you should be able to scroll down and see an **RP2040_Stamp** library entry. You can check that this has all worked by associating the correct stamp footprint to the **RP2040_Stamp** symbol you placed in the **Schematic Editor**, and then you can open and import the part into the PCB Editor. If you need a reminder on how to do those tasks, we covered them in the previous chapters.

Using the BMP280 module gives you an opportunity to learn how to make both a schematic symbol and a footprint. To do this you will create your own libraries to contain these parts, and others in the future. Let's begin with a custom schematic symbol. In the Schematic Editor, click the **Create, delete and edit symbols** tool button. This opens the Schematic Symbol Editor.

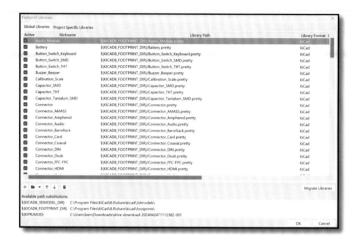

Figure 3-3 The **Footprint Libraries** dialogue where you can add or remove PCB footprint module libraries

In this new window, click **File** and then select **New Library** from the dropdown menu. You should now see a small dialogue box called **Add To Library Table**. In this box, you can select to add your new library to the **Global** table. This means any project in KiCad can access this library; alternatively, if you select **Project**, only this KiCad project can access that library. As the BMP280 is something you might use in other projects, make sure **Global** is highlighted and then click **OK**. You'll now be asked to give your library a name — it can be anything you want, so name it, making sure to leave the **.kicad_sym** part of the file name intact, and click **Save**.

You should now see your new symbol library name highlighted on the left-hand side of the Symbol Editor window. This means that this is the active library, so that when you select to create a new symbol, it will be stored in this library.

Click **File** and select **New Symbol** from the drop-down menu. You should see a **New Symbol** dialogue appear. Give your symbol a name that will be used in the library list — make it a useful name that reflects the part: **pi_bmp280**. Clicking **OK**, you should now see that a **U** has appeared in the Symbol Editor window and that the name of the symbol now appears in the active library in the list on the left-hand side of the screen. The name will have a * next to it, indicating that the symbol has not been saved.

In the Symbol Editor there are some familiar controls, such as the **F1** and **F2** to zoom in and out. Zoom out a little to give yourself some room and you can get started by adding some pins.

Click the **Add a pin** tool icon. In the dialogue, you can name the pin, give it a number, set its **Electrical type**, and change other settings if needed. For the first pin, name it **2-6V_in**, assign it pin number **1**, and set the electrical type to **Power input**. Continue to add pins 2, 3, 4, and 5, labelling them as you can see in **Figure 3-4**. As you place pins, notice that you can use the generic hot keys **M** for move and **R** for rotate, similar to the Schematic or PCB Editors. Once you have all your pins created, add a text label using the **Add a text item** tool. This lets you identify the symbol quickly when looking at a schematic.

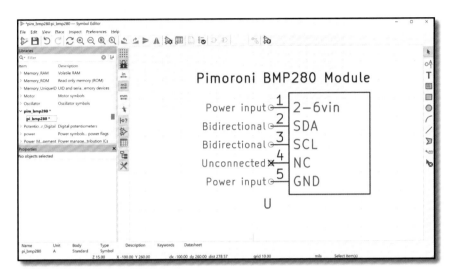

Figure 3-4   The Symbol Editor window can be used to create or edit schematic symbols

Finally, draw a bounding box around the schematic symbol so that everything is neatly grouped together. Click the **Add a rectangle** tool and draw a rectangle over your design. Click the **Save** icon in the top-left corner of the screen, and then close the Symbol Editor window. You can now go into the Schematic Editor and use the **Add a symbol** tool to find and add your first custom symbol.

Next, make a new footprint library and footprint for the Pimoroni BMP280 module. To begin, open the Footprint Editor (**Figure 3-5**). This is available from either the main KiCad project window or from the **Create, delete and edit footprints** tool icon in the PCB Editor. Similar to the Symbol Editor, once you have the Footprint Editor open, the first thing to do is to click **File** and then select **New Library** from the drop-down menu. Select to add a new library to the **Global** or **Project** table — select **Global** and name your library. Once the new library appears in the list, highlight it and then click **File** and select **New Footprint**. A dialogue appears and you can name your new footprint: Pim_BMP280_Module.

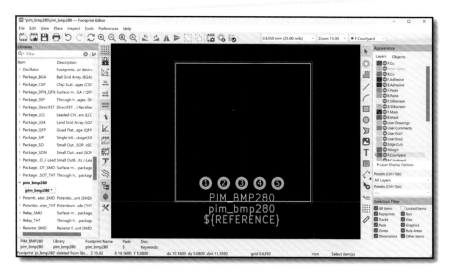

Figure 3-5   The Footprint Editor window can be used to create or edit component footprints

Quick Tip

If you use a schematic symbol or a footprint module from an external library, once it is saved in your KiCad project file, the symbol/footprint is stored in that project. If you moved to another machine with KiCad that didn't have the custom libraries added, you would be able to open the project as usual.

You also need click the drop-down menu and specify whether this is an **SMD** or **Through Hole** component. Select **Through Hole**, then add the pads and other parts of the footprint. Similar to the PCB Editor, you can set the grid resolution and, as the pins on the BMP280 module are spaced at a standard 2.54mm pitch, it is worth setting the grid to this initially to allow you to easily place the pins.

Next, click the **Add a pad** tool icon. You'll need to end up with five pads labelled 1 to 5 moving from left to right. The easiest way to do this is to count two grid spacings out from the centre datum line and then click to place pad 1. Notice that the **Add a pad** tool then increments the pad number so the next previewed pad is labelled **2**. Move one pad to the right of the pad you just made and click again. Continue until you have a neat row of five pads.

Reverting to the general **Select items** tool, hover over pad 1 and press the **E** key to open the **Pad Properties** dialogue (**Figure 3-6**). In this window you can change the geometry of the pad, the size of the hole, and many more options. We've found that increasing the pad size slightly from the default, and increasing the hole size, works well for soldering header pins between modules.

You may have your own preferences, but we edited each pad to be a 1.8mm circle with a 1mm hole.

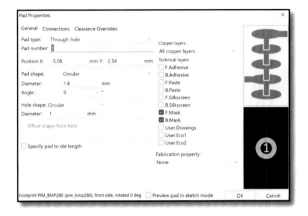

Figure 3-6   Editing the pad size and shape using the **Pad Properties** dialogue

With the pads created, you next need to add a silkscreen item that represents the physical area the module will occupy. Many components will have physical package dimensions listed in their datasheet, but that's not always the case with modules that are really designed for prototyping and breadboard use. In these cases, some investigation with a pair of callipers will help you get the dimensions of the package.

Measuring the Pimoroni BMP280, we realised a 19mm square area with the pin pads 2.54mm from the edge gives a slightly oversized and therefore safe margin for the module. Select the **F.Silkscreen** layer on the right-hand side of the screen and use the **Draw a rectangle** tool to draw and position it correctly.

Finally, add another square on the **F.Courtyard** layer that sits just slightly outside the silkscreen layer square you just created. You can achieve this by setting the grid to a very small spacing value such as 0.01mm and then drawing a rectangle away from the silkscreen rectangle to avoid it snapping.

Make the new square 19.4mm and then you can use the M key to move the square into position. This new square in the front courtyard layer provides a service called the 'DRC' or 'Design Rules Checker' with a boundary that shouldn't be overlapped. This means that later in the process, if a component is overlapping, this boundary on the PCB when you run the DRC will be highlighted as an issue.

USB On Board

You've now got nearly everything you need to make the Stamp and BMP280 board! You might have noticed a USB edge connector directly on the PCB in the main image. This is a cheap and cheerful approach to adding USB without adding any extra components, although you will need to add around 1mm of material to the back of the 1.6mm thick PCB in this area to actually make the USB connector fit. This highlights another way of using libraries and components from other, suitably licensed, KiCad projects. The USB Armory, which, in an early Mk1 version, used the USB edge connector on PCB approach.

USB Armory is an open source project — you can download the project repository here: **hsmag.cc/USBArmory**. Once downloaded, unzip the folder and use KiCad to navigate to the hardware folder, then mark-one, and then open the file **armory.kicad_pro**. Once open, move to the PCB Editor and select the **USB edge connector** footprint and press E. In the **Footprint Properties** window, click the **Edit Footprint** button. This should open the footprint in the Footprint Editor. You might get a warning that the footprint was made with an earlier version of KiCad, but saving the footprint in the editor should clear this warning.

You should see that the footprints pads and the silkscreen line (which doubles as a guide for the edge of the PCB board cut out) are now opened in the Footprint Editor. Across the top of the editor you should see a warning that you are currently only editing the footprint within the current project. You can use **File** then **Save As** to rename and save this footprint into your custom library that you created earlier. As that library is created on the Global table, this USB edge connector footprint is now available to use in any project. You should edit your version a little, labelling the pads more clearly, and save it to your library. Making a note of each pad's connectivity, you should repeat the earlier approach to create a custom symbol in the library to represent the USB connector in the schematic.

That's the components created. You should now be able to place them in your schematic, assign the footprint and then bring them into a PCB just as you have done previously.

Casting a net

You can see the complete schematic in **Figure 3-7**. In this circuit, there are more connections than you've used previously. While you could wire them up as you did before, the result would be quite a tangled web. To make things a little more readable, you can use *networks*, aka *nets*. This is where you give a name to a connection. Every connection to a particularly named net will be considered connected to each other.

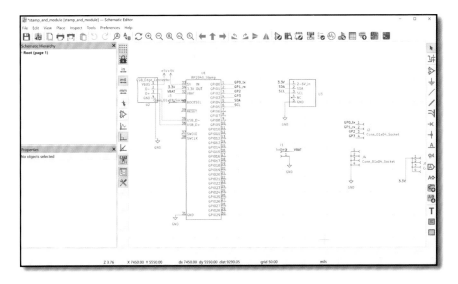

Figure 3-7 We've used both custom and generic connector symbols, as we did in the previous project

To add an item to a particular net, you need to add a label and connect it to this. Click the **Add a net label (L)** tool icon. You then click in the schematic and the **Label Properties** dialogue will appear. Into this you simply type the name of your net label, so, for example, type the letter **A**. You can change the font and size, but when you are happy, click the **OK** button and you can now place your **A** label into the schematic.

Notice that it has a small square connector. You can then connect a wire to the **A** label as you would to any other component using the **Add a wire** tool. You can recreate another **A** label using the same method to attach to the other end of your wireless connection, or you can copy and paste the original net label. If you have more complex descriptive net label names, it can be a good idea to use copy and paste (if you create a net label with a spelling mistake, it will not connect and may take you a while to discover the error). In this simple H-bridge example, you don't really need to use this technique but, in the next chapter with a more complex design, using net labels can really help to keep a schematic cleaner and more readable.

When you import the schematic into the PCB, any connections to the same net will be connected by lines in the rats nest.

Once you've made your schematic, you can import everything into a PCB. However, before you get started placing components, you need to create the board outline.

Carving a path

To create the board outline, you won't use the graphical tools in the PCB Editor but, rather, import an outline drawn in the free and open source Inkscape application (**Figure 3-8**). Whilst the included KiCad tools are excellent, Inkscape can offer some advantages when designing graphic components. For the example project, we drew a simple outline object for the board in Inkscape, saved it as an SVG file, and then imported it into KiCad.

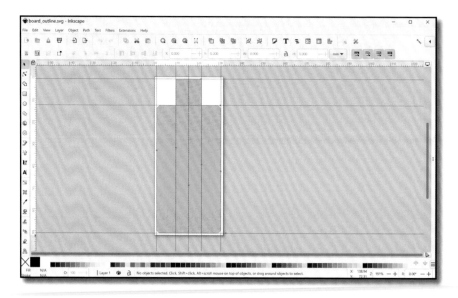

Figure 3-8   Using Inkscape to create an accurate graphic for the PCB edge-cut geometries

To do this, click **File** and then select **Import > Graphics…** from the drop-down menus. In the **Import Vector Graphics File** dialogue, you can navigate to the file and select the working layer to import to. Setting the graphic layer as **Edge.Cuts**, notice that you can also set an **Import Scale** value. In this instance we designed the board outline to be the correct size in Inkscape, so leave the import scale at 1.00. This function is useful, however, if you have an oversized graphic or logo to import to a silkscreen layer, like the HackSpace logo, we've reversed in Inkscape and then imported it onto the back silkscreen layer.

In the zone

Finally, another technique that we used on this board is to create flooded copper zones attached to a net label. This is an excellent way of automatically connecting groups of common connections. For example, many PCBs will have lots of connections to the GND net, so you can dedicate one layer (or part of one layer) to this net and simply cover it with copper (**Figure 3-9**).

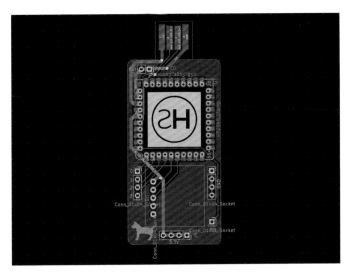

Figure 3-9 loading the board to connect all the GND connected pads

You can create complex systems with lots of different flooded areas with differing connectivity, but in this project you have just used one flood on the front copper surface that connects all pads attached to ground. To do this, once the board is laid out, use the **Add a filled zone** tool. Select this tool, then click to start drawing a fill area over your board design, ensuring you are on the correct layer. A dialogue appears and you can select the net connection from the list, so you should select **GND** (**Figure 3-10**). Leaving all the other settings at the default values, draw a rectangle over the board. Don't worry too much about accuracy as the flood will only appear inside the edge-cut geometry. Drawing three points of your rectangle you can then right-click and select **Close outline** from the drop-down menu. The rectangle (or other shape you drew) should now flood, and you should see that all the previously disconnected **GND** pads are now connected.

While the filled zone connects the entire GND net, you still have to wire the rest of it up manually. On the previous PCB, you could get away with just using a single side of the PCB because it was simple enough not to lead to any crossing wires. However, this is a bit more complex. PCBs almost always have more than one layer. Two is common for hobbyists, but you can go to 4 or even more for complex designs. By default, the PCB Editor has two layers, front and back. You can select the back layer by clicking on **B.Cu** in the right-hand pane. Now, when you draw a line, it will be on the back of the PCB, and will appear in a different colour (blue) in the PCB Editor.

Figure 3-10 Selecting the net to which the flooded copper zone will connect

It's all well and good having wires on both sides of the PCB, but how do you join them? Many of the components on the PCB are through-hole, and these create a link through the PCB and can be connected on either side. You can also add 'vias' — these are small holes that go through and link the front and back layers of the PCB.

To add a via, click on the **Vias** tool, then click at the place on the PCB where you want it. You can then connect wires to it on either side of the PCB and they'll be connected.

You might find that you need to place a track through the middle of your filled zone. If you do, this isn't a problem, you just need to use the **B** hotkey to recalculate the filled zone and it will skip out the area of the flood. It's possible this will mean that it misses a connection that should be in the net for the filled area; in this case, you may need to connect it in manually.

You can take a look at the finished project file for our solution to placing the components and wiring it up, but there are many ways of doing it and you may choose to do it differently.

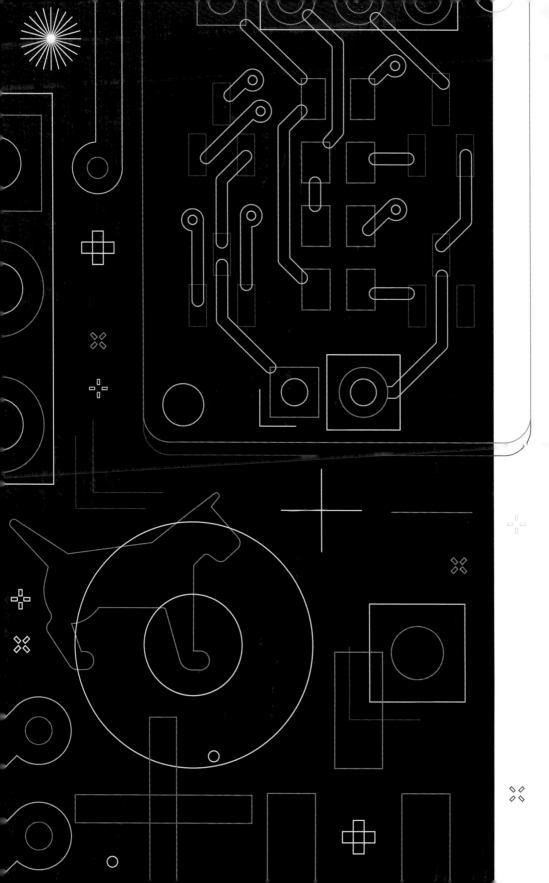

Chapter 4

Using a PCB assembly service

Create projects that can be fully assembled by popular PCBA companies

We covered a lot in the previous three chapters. If you worked through them all, you'll be at a point where you can create simple board designs. Now, you're going to do something a little more complex. You'll start with a small motor driver and build up to a development board based on the RP2040 microcontroller.

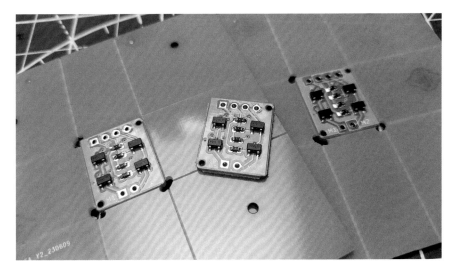

Figure 4-1   The PCB design successfully assembled by the PCBA service

The RP2040 chip itself comes in a Quad Flat No-leads 56 pin (QFN-56) package, and while that can be soldered at home using reflow or hotplate soldering techniques, that's difficult and time consuming. To avoid this, you are going to use a PCB assembly (PCBA) service.

There's a lot to look at in making an RP2040-based board, so in this article, you'll prepare a simpler design for manufacture to help you learn how to work with a PCBA.

Designing for PCBA adds complexity in that you have to create numerous files, define the PCB design, and choose and place components on the board. These components must be available to the PCB assembly house, which again adds complexity. The first few times you do this process, it'll seem like a lot of work compared to simply creating a PCB to assemble at home, but the payoff is that you get complete boards — no soldering necessary — and this is a significant time-saver as your PCBs become larger and more complex.

PCB Assembly

In earlier chapters, we suggested OSH Park for manufacturing PCBs — they make it easy by accepting KiCad PCB file uploads. If you want to use a PCBA service, it's likely you will need to plot *Gerber files* and files containing drilling information.

Gerber and drill files have some variables, and your fabrication service should give you some information regarding what they need. For example, JLCPCB has a page (**hsmag.cc/gerberdrillkicad**) that outlines the KiCad Gerber plotter settings that their service needs. It's a little out of date in that the screenshots are from KiCad 5.19, but you can find all the same options on the KiCad **Plot** dialogue. To access the latter, you click **File > Fabrication Outputs > Gerbers** from the PCB Editor.

Plotting Gerbers creates a bunch of files for different layers of your PCB design, so make sure that you create a folder for your Gerber files at the top of the **Plot** dialogue, or else they all end up mixed into the main root folder of your project. You can also create the drill file from the **Plot** dialogue after you have plotted your Gerber files. Clicking the **Generate drill files** button will launch a drill file dialogue. For JLCPCB, place your drill file into the same folder as you did your Gerber files and, finally, compress this folder into a zip file for upload to JLCPCB.

Other PCB fabrication houses will have guidance on what format they require. Don't worry too much if you upload something that doesn't work: it will give you an error and you can always ask the online chat service for help or guidance.

We'll look a bit more deeply at different fabrication houses and their requirements in Chapter 9, *Finding a PCB manufacturer*.

So, for this exploration of PCBA services, we've laid out a small design in KiCad for a motor driver circuit (H-bridge) using four N-channel MOSFETs.

We're not going to step through the board design, because we covered the approach in earlier parts of the book.

You can grab the project directly from **hsmag.cc/kicad_book_files**. That has the complete schematic and PCB. The project is in the **eg/ch04** subdirectory. You can follow along in this chapter to find out what we did to prepare them for automatic assembly.

You're going to use the popular and reasonably affordable JLCPCB assembly service to manufacture and assemble the boards. This means that you need to consider what parts you are going to use on the board, as each part needs to be available in the JLCPCB parts library. The first thing to do is to head over to the JLCPCB website (**jlcpcb.com**) and register for an account.

You can explore the JLCPCB parts library (**jlcpcb.com/parts**) using the search function and filters — you'll see parts' cost and availability. You can also advance-purchase components so that they are held in your own virtual warehouse ready to be used on your board designs in the future. This can be a very useful feature for parts critical to your project as there are few things more annoying that completing a tricky design only to find out that the parts are out of stock and may be for some time.

One thing of note is that available components fall into two distinct groupings: 'basic parts' and 'extended parts'. Basic parts will be added to your board at the price listed. So, if you are adding five basic part resistors at 0.07 dollars each, then they will cost 0.35 per board. However, if that part was listed as an extended part, then that part will still cost the same unit price, but there will be a one-time $3 setup cost of including that part in your project, as the part will have to be manually retrieved from storage and loaded into the pick-and-place machines.

As you peruse the JLCPCB parts library, if you spot a component that you are likely to use, make a note of the part number — these usually start with the letter 'C' and are listed on the main component landing page. For the small H-bridge board example, you'll only be interested in having JLCPCB add the SMD components, so choose some 10kΩ resistors (part number **C49122**), and the four MOSFETs (AO3400 chips, part number **C20917**) (see **Figure 4-2**).

You need to add these details to an extra field in **Symbol Properties** so that later, when you generate a bill of materials (BOM), these specific components will be identified. (To have JLCPCB ignore the through-hole components in a design, don't add this extra field to their schematic symbols. That way, they won't be included in the assembly process.)

To add these details, highlight a component in the Schematic Editor and press the **E** key to open the **Symbol Properties** dialogue. Click the + button,

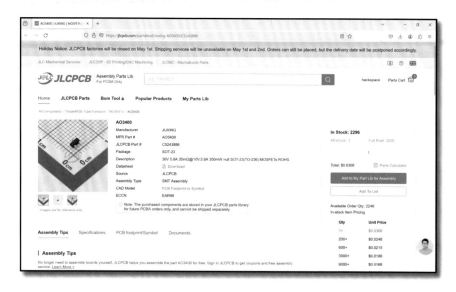

Figure 4-2   Searching for parts in the JLCPCB part library

which is labelled **Add field** when you hover over it (**Figure** 4-3). You should see a new field line appear. In the **Name** column, you need to label this field **LCSC**, (LCSC Electronics is a parts supplier and the parent company of JLCPCB) and then in the **Value** column, add the CXXXXX number you researched for that component.

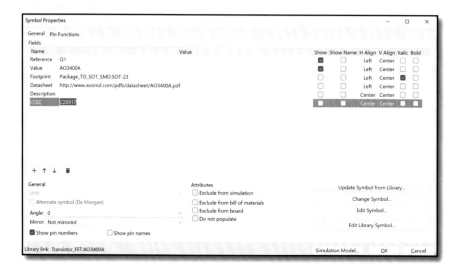

Figure 4-3   Adding an extra LCSC field to Symbol Properties enables a correct BOM to be created

You need to do this for every component you expect JLCPCB to add to your board; you can't just add this field to one 10kΩ resistor and expect it to add the same part for the others. For small projects, such as this example, you can do this manually for each schematic symbol. An alternate approach for larger projects is that you can edit the symbol at the library level or copy the symbol to a custom library with the LCSC field and number populated at the symbol level. This means that whenever you place that custom symbol in the schematic, you have the correct LCSC part number ready for the Bill Of Materials (BOM). Before creating the BOM, you still must assign the footprints to the symbols as you would for any PCB design.

Make sure you're in the Schematic Editor, then then click the **Generate a Bill of Materials** icon. In the dialogue that appears, click the **Edit** option tab in the upper left. You should now see something similar to **Figure 4-4** with your project components and details listed. In the left-hand side of the dialogue you can see a ticklist of visible fields which you can turn on and off. For a JL-CPCB BOM for assembly you will want four fields for each component: you will need a **Comment**, a **Designator**, a **Footprint**, and an **LCSC** for each component you wish to place. In most of the projects in this book, we've added a custom field to each component symbol for the LCSC numbers; notice that this custom field is included in the selectable field. If you uncheck everything from the list apart from **Value, Reference, Footprint**, and **LCSC**, you should see a preview spreadsheet like **Figure 4-5**.

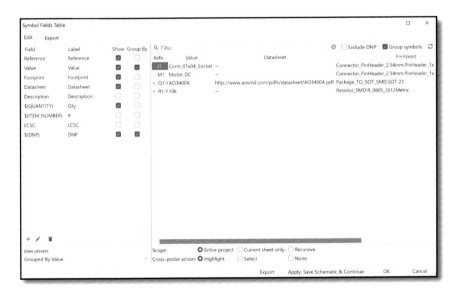

Figure 4-4   The Edit tab on the new KiCad 8 **BOM Editor** window.

It's likely that your column order doesn't match JLCPCB's required layout. If so, you can click the column title/label in the right-hand side of the dialogue

Figure 4-5 The correct columns containing the field data for the BOM

and drag it to another position. If the columns are ordered **Value**, **Reference**, **Footprint**, and **LCSC** from left to right, then this matches the JLCPCB BOM format. Finally, the column field titles won't quite match the JLCPCB requirements. In the left-hand side of the dialogue, click on the **Label** column to rename a field title. From left to right, JLCPCB requires a **Comment**, a **Designator**, a **Footprint**, and an **LCSC** — so you must make those adjustments.

Once everything looks correct, you can click back over to the **Export** tab in the left-hand dialogue and set a filename and set the filetype to CSV and click the **Export** button. Then, click **Apply, Save Schematic and Continue** to make sure this BOM format remains in this project.

In case you need it there are details about the BOM format required by JLCPCB here **hsmag.cc/jlcbom** and most PCB and assembly service will have similar pages detailing the BOM and other file formats needed.

Precise Placement

The final piece of the PCBA puzzle is to generate a *footprint position file*, also referred to as a *centroid file*. Like the BOM file, this is essentially a spreadsheet which contains details of each placed component, showing both coordinates and the rotational angle of the part.

To generate this, in the PCB Editor, select **File > Fabrication Outputs > Component Placement**. Make sure that your settings match the dialogue box shown in **Figure 4-6**. Note that if your project contains through-hole components that you want JLCPCB to include, you need to uncheck the **Include only SMD footprints** option and include the LCSC numbers for those through-hole parts in the BOM.

Don't include this component footprint POS file or the BOM file you generated earlier inside the zip file of Gerbers, as they are uploaded separately. Once you are ready, click the **Generate Position File** button. Notice that this will generate two POS files: one for the upper layer and one for the lower layer. As the design is single-sided, you are only interested in — and later, only need to upload — the upper layer file.

With the upper layer POS file generated, you need to make a few alterations to the spreadsheet column header titles for it to work properly for JLCPCB. Open the file you have generated in your spreadsheet program. You can use LibreOffice Calc, Microsoft Excel, or Google Docs.

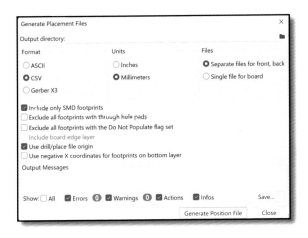

Figure 4-6   The **Generate Placement Files** dialogue

Next, make the following changes: change the title of the first column, **Ref**, to **Designator**; PosX and PosY to **Mid X** and **Mid Y**; Rot to **Rotation**, and finally, **Side** to **Layer** (Figure 4-7). Save the POS file with these alterations — you now have everything you need to upload to the JLCPCB service.

With everything ready, head to the JLCPCB website. Click on the **Standard PCB/PCBA** tab to upload the Gerber zip file. After a short upload, you should see a render of the upper and lower sides of your board in the preview window (**Figure 4-8**). You can make changes to the board type, material, thickness, and more on this initial page. However, apart from changing the colour of the board to yellow (this is purely decorative — we wanted something a bit brighter than the standard green), we left everything at the default setting.

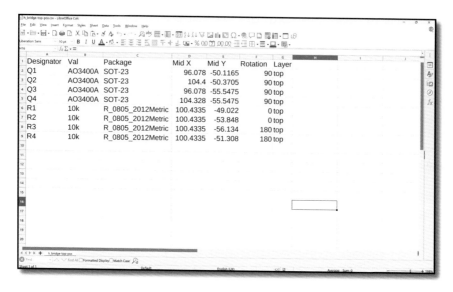

Figure 4-7   Using LibreCalc to edit the generated positional file column titles

Figure 4-8   If the Gerber ZIP file uploads correctly, you'll be rewarded with the first of many preview images of your board

At the bottom of the page, you can click a button to add/expand the PCBA services. In **Figure 4-9**, you can see that we have opted to have the top side assembled only (as you only have components on this side). Depending on whether you've selected Economic or Standard Assembly, you'll want to high-light select either **Edge Rails/Fiducials** or **Tooling holes** as **Added by JL-CPCB** option highlighted. This indicates that for these very small boards, JLCPCB services will create any needed panel layouts. With all that selected, click **Next**.

You'll get another larger preview of the PCB layout generated from the Gerber uploads. Check it carefully and then click **Next** to move to the next tab. This will look like **Figure 4-9**. Click the **Add BOM File** button and upload the BOM file you created earlier, then click the **Add CPL File** button and upload the top layer CSV positional file you made and edited earlier. Clicking **Next**, these will be uploaded and processed, which may take a few minutes, and you should see a render with the board and the components placed.

Figure 4-9   The preliminary choices for the PCBA service

Spin me around

Often, at this point, you will find that components are not rotated correctly on the footprints. There are two ways to correct this, if your components are at standard angles, you can click to highlight a component in the render image and, when they are highlighted, use the rotation tools above the PCB render in the JLCPCB web page (**Figure 4-10**). You can continue to do this until your design looks correct. Clicking **Next** will save these orientations and take you to the **Add to Basket** ordering page. Whilst this is a fine approach, another option is to open the positional file you created and uploaded and then edit the rotational value of the components that have appeared incorrectly in the render. In our experience, either way is fine, and the JLCPCB engineers will usually question if a component isn't sitting on a footprint correctly (though it's best not to rely on this).

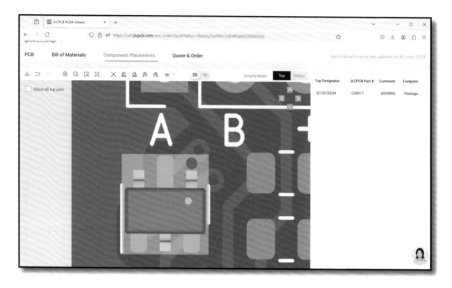

Figure 4-10 Correcting footprint rotational position can be done in-browser as part of the JLCPCB order process, or you can edit your positional file offline to create correct values

All that's left to do is to add the order to your shopping basket and pay. Once the order is confirmed and paid, you get regular updates on the order listing and, if it's a complex design, it's worth checking back into your account four to six working hours after placing the order to check the analysis regarding component placing in the order history details.

Once everything is ordered, all you have to do is wait. The exact times vary, but it's often quite quick. We started this simple H-bridge motor driver design in KiCad and had the assembled PCBs (**Figure 4-1**, shown earlier in this chapter) in our hand eight days later.

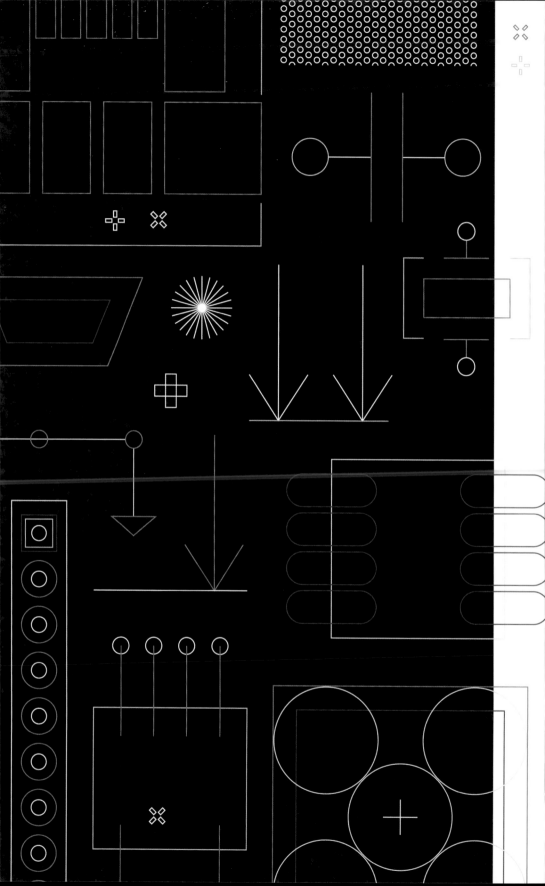

Chapter 5

Designing an RP2040 board

Get your custom microcontroller board made for you

In the previous chapters, you worked through the basics of making PCB boards from schematic to board layout, and you learned a variety of skills and approaches along the way. You built on this by exploring how to use KiCad and a PCB assembly (PCBA) service to not only have the PCB manufactured, but to also be populated with components and supplied to you fully assembled (**Figure 5-1**).

Figure 5-1 A fully assembled and functional RP2040-based board

RP2040 is a great target for PCB projects that will be assembled using industrial approaches. The bolder amongst us might successfully be able to solder up the QFN 56 package at home, but PCBA makes everything a lot easier.

Many commercially available boards that use the RP2040 are manufactured using PCBs with four or more layers. Of course, it's possible to do this in KiCad, but for many of us, four-layer boards can be difficult to debug or correct if something goes wrong. Fortunately, two layers is enough to get a microcontroller running and break out the features you need.

When RP2040 was released, so too was a stack of excellent documentation, including the surprisingly readable *Hardware design with RP2040* (**Figure** 5-2). This document shows an example of a minimal RP2040-based board, a two-layer design, and then describes various aspects of the design and considerations. There is even a KiCad project file for the design. It's a great idea to download the project file and review it (**Figure** 5-3). In this chapter, you'll replicate this board design, but you'll start from a blank project rather than using the one supplied.

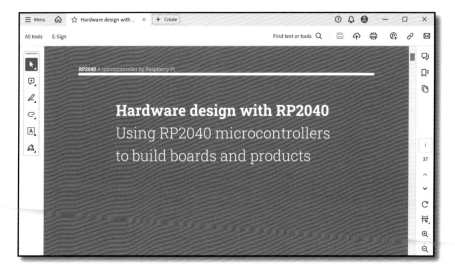

Figure 5-2   The excellent *Hardware design with RP2040* documentation from Raspberry Pi

Starting from scratch means that it's easier to adapt this project to your own needs. You'll also tweak the project because currently, JLCPCB doesn't supply some of the exact components used on the Raspberry Pi example. Also, the Raspberry Pi example was built in a previous version of KiCad and uses in-house schematic symbols for the RP2040. Since then, the RP2040 has become a standard library component within KiCad.

Some of the connections on the RP2040 schematic symbol are already made at a symbol level (like common power pin connections), so it makes sense to use the built-in KiCad symbols and footprints. With that all said, this chapter won't step through every stage of designing this board as you've learnt the basics in previous chapters, so you'll learn things that are specific to creat-

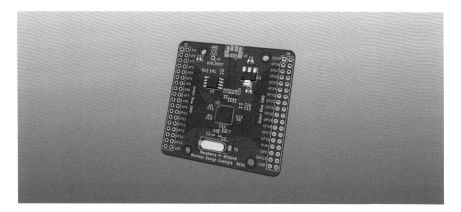

Figure 5-3   The *Hardware design with RP2040* documentation also includes a KiCad example project

ing an RP2040 board. You can download the KiCad files for this project from **hsmag.cc/kicad_book_files** in the **eg/ch05** subdirectory.

This chapter's project emulates both the schematic layout and the PCB layout of the minimal design example. This is partly because the PCB layout has neatly solved a lot of the layout complexity, but it also allows you to compare the two projects as you build your own.

Planning ahead: components and footprints

When designing for JLCPCB, even at the schematic level, you need to be thinking about what component and footprint you will be using. For the RP2040 board, we wanted to use a surface-mount micro USB-B socket where all the USB chassis and ground points are on the upper layer — this means it could be assembled as an SMD component. As such, apart from the pin headers/sockets, the units would be fully assembled by JLCPCB.

Looking through the LCSC component library (**lcsc.com**), we opted for the **C132560** component. JLCPCB often supplies schematic and footprint symbols for EasyEDA. Fortunately, you can import this into KiCad (this is a new feature in KiCad 8, so make sure you're running the most up-to-date version).

To get the files, you'll need to set up an EasyEDA account (**easyeda.com**). Click the link in the LCSC component library that says **PCB Footprint or Symbol** on the component page on the JLCPCB library website, then click the **Free Trial** button underneath the pop-up window that shows the component symbol and footprint.

After setting up a login, you should see a browser-based EasyEDA project with just the footprint of the component loaded.

There are numerous tabs in EasyEDA and you should be able to find and swap between two tabs, one with the component schematic symbol and one with the component footprint. With the component tab selected and in view, click **File > Save As Document Save As (local)** (**Figure 5-4**). When clicked, a file with the filetype EFOO should download. Whilst you are here, swap tabs in EasyEDA so you see the schematic component, then repeat the same **File > Save As Document Save As (local)** process and you should download a file with the filetype ELIBZ.

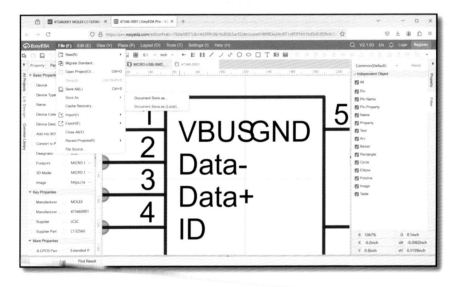

Figure 5-4 Use the local save as option in EasyEDA to download schematic symbols or component footprints

Moving to KiCad in the main project view, click to open the Footprint Editor. Next, use **File > Import > Footprint...** and navigate to the downloaded EFOO file, **Figure 5-5**. Select the file and the footprint will be automatically imported. Next, save the footprint to a library or custom library in the usual way using the **File > Save As** dialogue.

The process is the same for schematic symbols. Open the Symbol Editor and use the import function to import your downloaded ELIBZ file. If this is the first time you have opened the Symbol Editor or the Footprint Editor you will need to select a working library before being able to import. However, you can save the imported item to any library you choose.

Now you've done this, you can use the symbol and footprint just as any others.

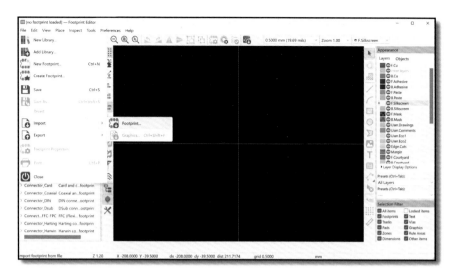

Figure 5-5   You can import the EasyEDA footprint into the KiCad footprint editor

With the USB symbol (and footprint) sourced, add the **RP2040** component from the KiCad library. Next, add resistors and a pair of labels to connect the D+ and D- to the correct pins on RP2040. The documentation states that you need to create the traces for these connections with accurate dimensions and clearances. Do this by assigning a custom net class to the connection, or *net*, so when you draw these traces in the PCB Editor, they'll be created correctly.

Quick Tip

Do take care when creating custom footprints to ensure that the component pin number is compatible with your schematic symbol and vice versa.

In the **Schematic Setup** dialogue in the Schematic Editor, click the + button in the uppermost **Net Classes** window, add a new net class, and give it a name. Note that local net class names within a project should start with a / — use **/USB_lines**. Select a wire segment that connects the **RP2040 D+** or **D-** pin out to the **USB_D+** or **USB_D-** label that you added, and then right-click. Select **Assign Netclass** from the drop-down menu. In the **Add Netclass Assignment** dialogue box, you should see the selected label **USB_D+** and, to the right of it, a drop-down menu to select the net class — this is currently **Default**, but if you click down, you should be able to see the **USB_lines** net class that you added earlier. Select this, then close the dialogue box. Repeat this for the **USB_D-** label as shown in **Figure 5-6**.

After opening the PCB Editor, use the **Board Setup** tool (in the same positions as the **Schematic Setup** tool) to adjust the net class variables. This includes

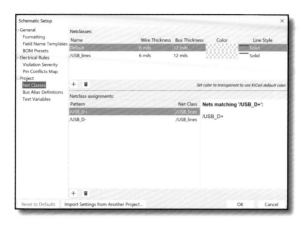

Figure 5-6   Setting up and assigning a net class

the track width, track clearance, via size, and more (**Figure** 5-7). You'll need to set the track width and the track clearance to create the track lines that you need for the USB lines — this information on track geometry was taken directly from the *Hardware Design with RP2040* documentation (see section 2.4.1, *USB*).

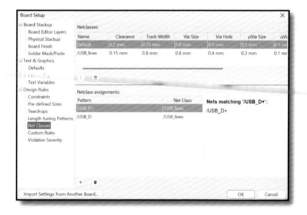

Figure 5-7   Setting the net class variables for the USB lines in the Board Setup dialogue in the PCB Editor

We laid out the power regulator and found a similar device to the minimal design example; however, the JLC component **C26537** had an extra pin and was in the **SOT-223-3** package. We have a matching KiCad library footprint, but it was easier to just make a quick custom schematic symbol to ensure that the pin numbering was correct and matching (**Figure 5-8**).

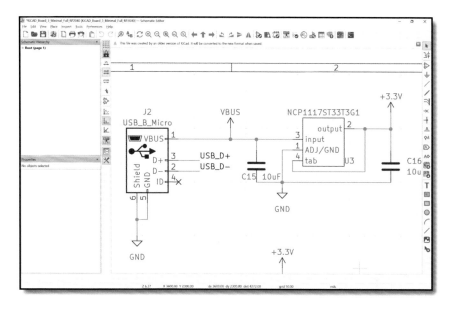

Figure 5-8   Setting up the power regulator is simple once the target JLCPCB components are identified

Trace widths

Often you will want to use thinner and smaller traces simply to save space and allow you to route out complex chips without the board becoming too big. PCB fabrication services will usually have the minimum trace size that they can handle listed on their site somewhere. So why don't you always just use the smallest trace size your board maker can supply? Well, obviously a trace is like a wire and can only pass a specific amount of current before the trace starts to get hot, and in worse cases, melt or even set on fire. Most PCB designs will have some parts of the board that are very low current with other areas having higher current needs. A rule of thumb for RP2040 boards is that traces breaking out GPIO can be pretty thin as the GPIO have an absolute current limit of 50mA, but more typically you want to only draw up to 12mA. Compare this to the power supply lines coming from the USB connector where these might be drawing up to an amp. To support these different uses, you need different trace widths.

Calculating trace widths and current handling can get a little complex, but there are lots of calculators online that can help. You need to know (or estimate) the height of a trace on a PCB (a PCB house that uses more copper in traces to make a taller trace might be able to use thinner trace designs an achieve the same current handling characteristic).

For example, the standard copper weight for boards from JLCPCB is 1oz. This results in a trace height of just a shade under 1.4mils (where mils are 1/1000th of an inch). It's more common for copper weights and traces to be calculated in imperial units, but you can convert it over to mm if needed.

? Tooling Holes

When ordering assembled PCBs from JLCPCB, you might need to consider tooling holes. These are small holes placed in your design layout that JLCPCB machines use to locate and hold the boards when they are being assembled. You can add these yourself, or you can omit them, knowing that JLCPCB will place them for you. It's fair to say that JLCPCB engineers will place the holes sensibly and won't plonk one through a trace or in the middle of a component footprint; however, you might want to manually add them into your design so that you decide where they are finally placed.

The rules are that a minimum of two, preferably three, holes should be placed in the PCB design, and they should be placed at opposite corners — as far apart as they can practically be. The holes should be 1.152mm diameter circular non-plated holes with a 0.148mm solder mask expansion. we found that the easiest way to create these tooling holes is to create a custom component in the KiCad libraries that you can drag into and place in the design.

To make a tooling hole component footprint in the **Footprint Editor**, you need to create a new footprint and then add a single pad. Selecting the pad, press the **E** key to enter the **Pad Properties** dialogue. On the first **General** tab, set the pad as **NPTH, Mechanical** in the pad type — **NPTH** stands for Non-Plated Through-Hole. Staying on the **General** tab, make sure the pad shape is set to circular and the diameter is 1.152mm. Finally, click onto the second tab in the **Pad Properties** dialogue called **Clearance Overrides and Settings**. On this tab, set the solder mask expansion to 0.148mm. Save this as a footprint and you can place them when needed into JLCPCB-oriented designs.

As you're not aiming for any particular use case with this example, you can simply break out all the pins to a pair of headers that you will place on each side of the board. With the schematic largely complete, you can move over to the PCB Editor to begin the layout and routing.

You can use an online calculator to determine the required trace width. You can find an excellent calculator at **hsmag.cc/tracewidth**. As a simple example, set the current value to 1 amp and the copper thickness to 1 **oz/ft^2**. For a simple value, ignore the optional inputs and move straight to reading the results for an internal layer. This says a value of 30.8mils. If you convert this into mm (by multiplying by 0.0254) you get roughly 0.78mm. If you use a trace larger than this, for example 1mm, then you know the trace is overspecified.

Finally on trace widths, there is another factor that you may need to consider: Impedance. For PCBs running at lower frequencies including all the projects in this book, it isn't an issue at all. However, if you move to higher frequency designs, it can cause problems where the trace can begin to create magnetic fields and electric charges around the traces which can interfere with the circuit characteristics. It's a long and involved subject that we'll sidestep entirely as it's not relevant to the designs in this book. However, if you would like to know more, or are concerned about it causing problems in your design, you can find a good introduction at **hsmag.cc/impedance**.

Crystal conundrum

When looking up the components that the minimal design example project used, one area that challenged us was the crystal; we found no similar device available on the JLCPCB component library (**Figure 5-9**). Using a consummate hardware hacker's approach, we looked into what components other open source RP2040-based designs used, making sure to limit the list to projects we knew worked well. Having used Solder Party's Stamp earlier in this book, we checked out its design. It looked simple and straightforward with just two 12pF capacitors, and on checking JLCPCB, the crystal part was available as part number **C521567**. To use this, you'd need to again convert the footprint from EasyEDA.

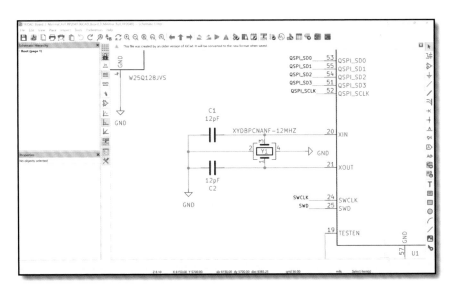

Figure 5-9 Due to JLCPCB not stocking the crystal that the RP2040 documentation recommends, we needed to rethink this part of the circuit

Laying out the decoupling capacitors in the schematic is straightforward, and you should take advantage of the KiCad text tool to add notes, occasionally acting as reminders for important information. In the *Hardware design with RP2040* documentation, for example, it shows that the 1uF capacitors should be placed close to pins 44 and 45 on the RP2040, so it makes sense to add a schematic note as a reminder (**Figure 5-10**).

The flash memory chip used in the minimal design example is available at JL-CPCB and, as such, you should go with the same design. In the *Hardware design with RP2040* documentation, they have added a footprint for an optional pull-up resistor but found, with this chip, that it wasn't needed, so you can omit that part of the design. The important thing about the flash chip and the associated traces on the PCB is that it all sits over a continuous ground plane. This means you must think carefully about routing the traces, especially on a two-layer board, so use this project or the Raspberry Pi project as a starting point.

One interesting aspect of the board design, that we haven't looked at yet, is that it has numerous different voltages which each have their own copper flood zones. Making these is like how you make any other copper flood, as in previous boards, but you need to set a priority value so that the different floods know how to flood separately. So, on the top layer of the board there is a general 3V3 flood, a flood connected to the **VBUS**, which is the 5V input from the USB to the voltage regulator, and there is a small 1V1 zone inside the footprint of the RP2040. Notice in **Figure 5-11** that we have assigned the 1V1 flooded area priority 2, the **VBUS** area priority 1, and the general 3V3 priority 0. This essentially shows that they are separate areas.

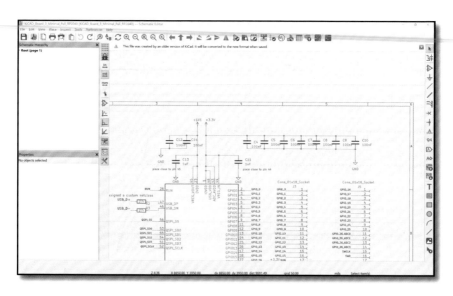

Figure 5-10 Laying out the decoupling capacitors in the Schematic Editor is largely straightforward

We decided that for the resistors and most of the decoupling capacitors, we would go for 0402 size packages (we're using the imperial sizes because these are most common in the hobbyist community). Again, we wouldn't make this choice if we were planning to assemble this board by hand, but with the assembly engineers and robots doing this fine work, you might as well use the tiny packages. It's less common for the very tiny packages of capacitor to be of accurate value as they increase in capacitance. So, although JLCPCB does offer components that claim 10uF in 0402 packages, you should go with a more common, larger 1206 variant.

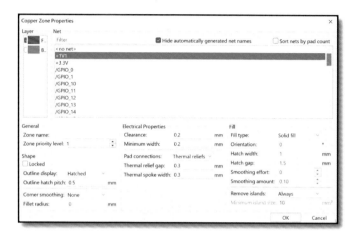

Figure 5-11   Setting up different priorities to allow different zones to coexist

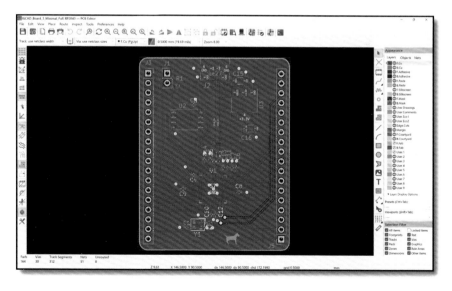

Figure 5-12   The completed layout in the PCB Editor

That covers everything unique to this project. If you've worked through the previous chapters, you can apply that experience to this project. Smaller projects like the H-Bridge design in the last chapter will help you get used to JL-CPCB-specific processes. Reading *Hardware design with RP2040* a couple of times before tinkering with an RP2040 board is time well spent. Finally, be warned, having fully assembled and hopefully functional boards delivered to your door is highly addictive!

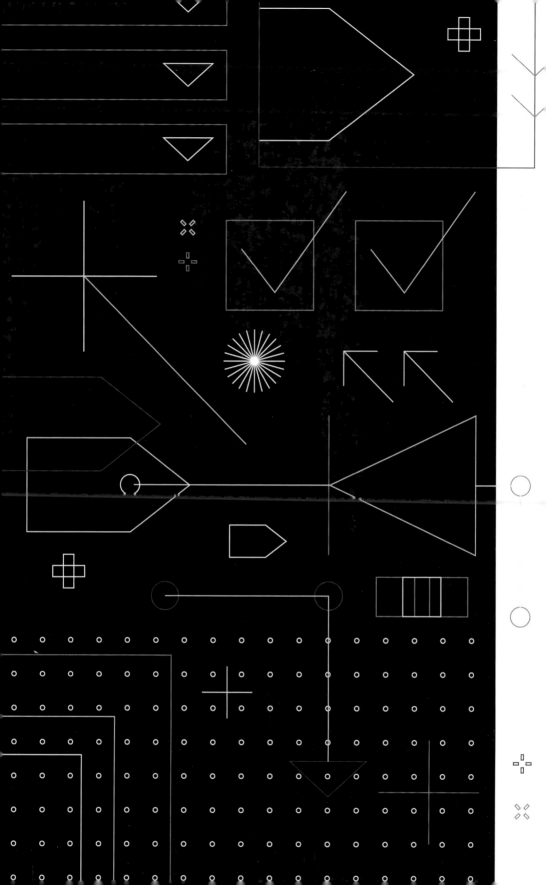

Chapter 6

Schematic organisation

Get your sheets in order before starting a larger project

In the last chapter, you laid out your largest project so far — a minimal RP2040 board and, in the next chapter, you'll learn how to add to this design. However, the schematic is already quite full and, if you're not careful, the whole thing could become unmanageable. This chapter will show how you can keep things clean, tidy, and easy to work with.

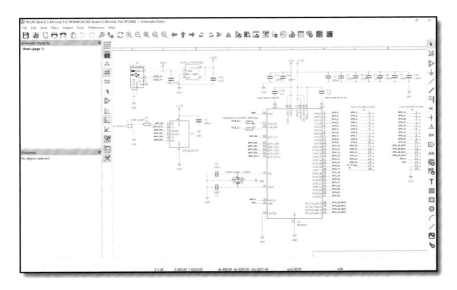

Figure 6-1   The schematic from the RP2040 minimal example project

If you view last chapter's project, and jump into the Schematic Editor, it looks quite cramped (**Figure 6-1**). One of the first things you can do is to simply increase the schematic size. Navigate to **File > Page Settings** and, in that dialogue, you can swap the page size, currently at **A4**. Changing this to **A3** will give you plenty more room.

While you are in this **Page Settings** dialogue, add some detail to the **Title Block** fields. This is the lower right-hand corner collection of text boxes that list the **title, revision issue date**, and more. Clear titles and revision labels and dates are useful if you plan to publish your project or use the schematic as part of documentation (**Figure 6-2**).

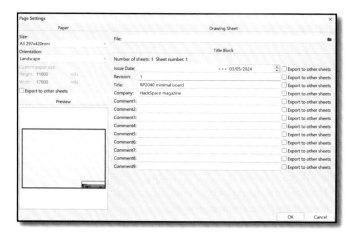

Figure 6-2 Increasing the schematic page size and adding clear labelling is a good start to keeping organised

With the schematic size increased and neatly titled, let's further organise your schematic by creating graphic boxes around different subsections of a project's circuits with the **Add a rectangle** tool. First, use the selection tool to select and move parts of the schematic into a position where you can completely draw a rectangle around them.

Quick Tip

If you can't select a section of schematic completely using the selection box tool, select as much as you can, then press and hold the **CTRL** key whilst clicking any missed components or wires.

Select the **Add a rectangle** tool (in the lower right-hand side). Click the start position (a corner of your rectangle) and then move the pointer to the opposite corner position you desire. Once the rectangle is drawn, click it with the selection tool, and you can either grab the anchor points to change its dimensions, or press **M** on the keyboard to move the rectangle as you would any other object in the Schematic Editor.

With a section of your schematic now encapsulated in a rectangle, use the **Add text** tool to create and then place labels into the boxes (**Figure 6-3**). You did this in the previous chapter's RP2040 minimal example project when you added notes to place capacitors close to certain points. If you want to bring a little of your own house style to schematics, you can use any font for the text, rather than the default KiCad fonts.

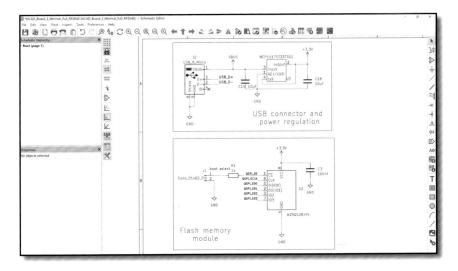

Figure 6-3 Adding boxes and text labels can help organise sections of complex schematics

Roving Robot

We're going to use last chapter's RP2040 example into new projects in future chapters. You'll want to keep the original RP2040 board project, so you need to make a copy of the project to develop into a robot rover platform.

To make a copy of the original project, go to **File > Save As,** then select or create a new folder to save the project to. Type a new name for the copy in the **Name** dialogue box, then click **Save**. Although not completely necessary, it's not a bad idea to then open the new project folder in a file browser and delete any unneeded files for the new project. For example, if you have a Gerber file folder, these probably won't be relevant to the new project. Any backup files can also be removed as you would probably return to the original source project, if required, rather than use a backup from this new branched project.

The first new project is a wheeled robot. The basic premise for the robot is that it is going to have four wheels, all of which are driven by N20-style motors. This gives you options down the line in that it can run with normal style wheels and tyres, but would also work with *Mecanum wheels,* which need to be driven individually to create the interesting sideways and diagonal motions Mecanum wheels are known for.

A primary part of the design is to add four motor driver circuits to the RP2040 board — one for each wheel. You could place the four motor driver circuits into the A3 schematic and wire them using either labels or direct wiring. However, another way you can add content to the schematic is with *hierarchical sheets*. A hierarchical sheet is a sub-sheet that exists in the schematic below the main page, into which you can insert designs that can be connected to the main top-layer schematic page.

A splash of colour

You can add images to schematics and, while you can place images anywhere on any page, a common use case for this is to add a company logo to the **Title Block** area. The Schematic Editor supports a wide range of image formats, including PNG, JPG, TIFF, and BMP. You can rescale the image as you import it. This is useful as you can import a larger, higher DPI image and then scale it down to fit the box.

To add your image to the schematic, click the **Add a bitmap tool** icon on the lower right-hand side of the screen, and navigate to your chosen image file. Once the object is imported, you can move it and scale it whilst maintaining its aspect ratio by dragging the corners of the image bounding box.

To create a hierarchical sheet, you can use the **Add a hierarchical sheet** tool icon found in the lower right-hand area of the Schematic Editor, or you can press the S key on your keyboard. Next, click and drag to create a rectangle in the schematic. When you click to finish the rectangle, the **Sheet properties** dialogue appears. Into this you can put a sheet name, which will be displayed in the rectangle on the main schematic page. Name the first sheet **L9110_Motor_Driver_1**. Under this is an input box for a sheet file name. It will currently be populated with **untitled.kicad_sch**. This is the name of a separate file that will be created for this hierarchical sheet. Change this to something appropriate, such as **L9110_Motor_Driver_1.kicad_sch**.

Notice that there are checkboxes labelled **show** and, by default, they will be ticked (**Figure 6-4**). This means that, in the main schematic sheet, the hierarchical sheet rectangle will appear with both pieces of information listed. It is advisable to show one, or both, of these pieces of information or else you must navigate into the hierarchical sheet to see what it contains, which can be confusing if you have multiple hierarchical sheets. Similarly, notice that you can also play with the appearance of the hierarchical sheet rectangle, increasing or decreasing line width, background, and border colours. This might seem frivolous, but when you work in designs with many of hierarchical sheets, being able to differentiate them by colour can help you work quicker.

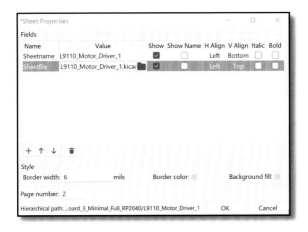

Figure 6-4 Adding a name for a new hierarchical sheet and a file name for the new hierarchical sheet schematic file

To enter your new hierarchical sheet from the main schematic page, you can either right-click on the rectangle (**Figure 6-5**) and select **Enter sheet**, or you can double-click on the hierarchical sheet rectangle. You should be met with a brand-new empty schematic.

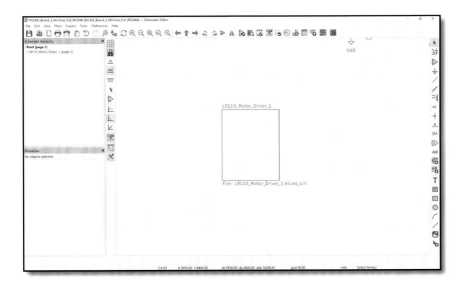

Figure 6-5 The empty hierarchical sheet placed in the original schematic

You can now begin to place components and create your circuit in the hierarchical sheet. The first thing to note is that anything connected to a global label or net will automatically be connected to those points globally. If, for example, you add a component and connect it to **GND** and to **VCC**, they will automatically be connected to those points in the top-layer schematic and will be connected when you import those components and connectivity to the PCB Editor.

If you did place and connect such a component in a hierarchical sheet, when you return to the main sheet you won't see any connections coming out of the hierarchical sheet rectangle. For general ground and power connections this might be OK, but it may get confusing if you use this approach for all your connectivity. A common way of making connections into and out of hierarchical sheets is to use *hierarchical labels*.

> **Quick Tip**
>
> As a hierarchical sheet is a separate schematic file, you can use **File>Page Settings** to set the title, page size, and other details, as you did earlier.

Digging deep

To place a hierarchical label, you can either click the **Add a hierarchical label** tool icon or press H on your keyboard. A **Hierarchical Label Properties** dialogue will appear. Insert a name into the **Label** field. You will now have a label you can place, move, and rotate and connect into your design.

In **Figure 6-6**, we have made the hierarchical label **test_hierarchic** and connected it to the anode end of an LED component symbol. We've also placed a global label test to another LED component. Moving out of the hierarchical sheet is simple: you can either right-click and select **Leave sheet** or you can hold **ALT** and then tap the **Backspace** key.

Even after creating a hierarchical label inside a hierarchical sheet, you won't see that connection until you right-click over the hierarchical sheet rectangle and click **Import Sheet Pin** from the drop-down menu. You should now see any hierarchical pins appear aligned with the edge of the rectangle. You can move these pins to any position around the rectangle, and you can connect and wire to these labels as you would any other symbol.

In **Figure 6-7** you can see the **test_hierarchic** label in the sheet rectangle, and it is wired to **VBUS**. Note that you can't see the global label **test**, but that point inside the hierarchical sheet will be connected to any other connections with that global label anywhere in the project schematic. Inside this hierarchical sheet, there is a connection to a **GND** symbol and, as such, that point is connected to the project's global **GND** label.

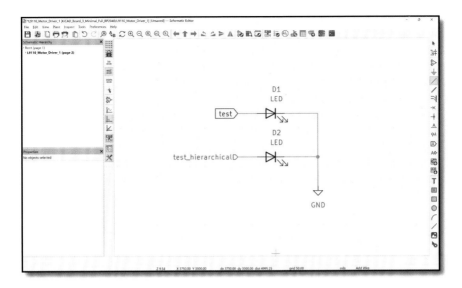

Figure 6-6   Inside a hierarchical sheet with a design using a hierarchical label and a global label

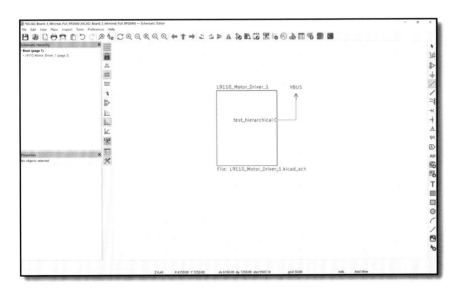

Figure 6-7   The test hierarchical sheet object viewed from the main schematic. The hierarchical pin has been connected to VBUS

For the robot rover design, let's keep it simple and affordable and use the L9110 motor driver IC as it's adequate for the N20 motors, affordable, and there is a large amount of stock available on JLCPCB. The circuit around the L9110 is straightforward (**Figure 6-8**), with a couple of pull-up resistors and a decoupling capacitor across the motor outputs. When we created a hierarchical sheet for the motor driver circuit, we used hierarchical labels to create the two signal inputs. We could have opted to have the motor outputs as hierarchical labels, but it was cleaner to add the motor output connector symbols inside the hierarchical sheet, avoiding more clutter on the main schematic.

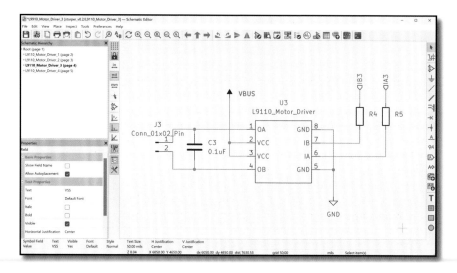

Figure 6-8 The L9110 circuit for each motor. Note the hierarchical labels used on the IB and IA pins

One of the great benefits of using hierarchical sheets is that you can quickly copy them (or their contents) to create multiples of similar modules, either within the same project or into other projects. In the robot rover project, we have simply added three more hierarchical sheets and named them sequentially as **Motor_Channel_1**, **Motor_Channel_2**, **Motor_Channel_3**, and **Motor_Channel_4**.

We then copied and pasted the contents of **Motor_Channel_1** into each different hierarchical sheet. Each time we copied the L9110 circuit, we relabelled the hierarchical pins so that each motor channel was unique (**Figure 6-9**). You can then use the **Import Sheet Pin** function used earlier on each of the motor channel hierarchical sheets, and you are ready to wire the pins to your chosen GPIO pins on the RP2040 — either directly or using labels to again keep the main schematic clean and tidy.

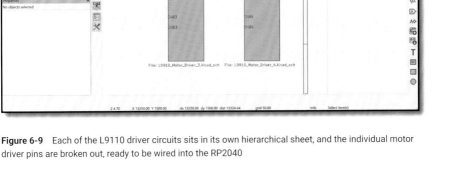

Figure 6-9 Each of the L9110 driver circuits sits in its own hierarchical sheet, and the individual motor driver pins are broken out, ready to be wired into the RP2040

To help you understand hierarchical sheets, you should look through some open-source hardware projects that have used KiCad. You'll certainly find lots of different approaches to organisation and project management.

Large projects

One thing about KiCad is that you can use many different approaches, depending on your needs or your way of thinking. One such approach is to use a flat hierarchy for your schematic layout. This essentially turns the first schematic sheet into a kind of holding sheet where all the hierarchical sheets are held. You can then, in each sheet, use general and global labels so you don't need to draw any connectivity on the main topmost schematic.

Although this might momentarily confuse anyone else opening your project, it's a brilliant way to organise a particularly large project into specific sections. As each hierarchical sheet is an individual schematic file, you can simply work on each file individually as needed in development.

If you copy and paste a hierarchical sheet rectangle in the main schematic, it will automatically add and increment a number to the name of the sheet. So, **sheet** would become **sheet1**, then **sheet2**. If you created a first sheet called **Something_1** and copied and pasted it, the clone would increment to **Something_2**.

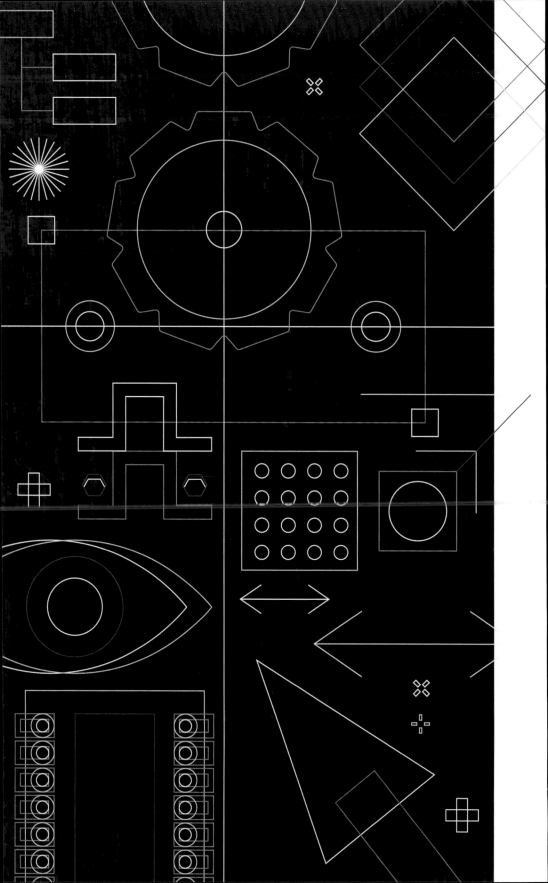

Chapter 7

Get the right shape

When your PCBs are part of the structure, they need to be accurate

It's increasingly common for projects to incorporate PCBs as a mechanical part of an object. In the last chapter, you looked at hierarchical sheets and laid out a motor driving circuit that you could copy and paste to add motor drivers to a project. In this part, you are going to create a simple robot rover that we call 'stoRPer'. StoRPer is a tongue-in-cheek reference to a favourite childhood toy from the 1980s: the 'Stomper'.

Figure 7-1   The stoRPer robot prototype using the PCB as its main chassis component

The Stomper, by toy company Schaper, was the first ever four-wheel drive electric toy car. Despite no form of remote control, they were great fun to try and build obstacle courses for, or to test on steep gradients.

We wanted the reasonable torque and the four-wheel drive aspects of the Stomper but with the addition of a Raspberry Pi Pico to make it a more interesting and controllable platform — so, 'stoRPer' it is. It's designed with all-wheel drive (AWD) so that we can work with Meccanum wheels. You can find the files for this project at **https://hsmag.cc/stoRPerGit**.

You will use a Pico as a module on this build and focus on using the main PCB as a mechanical part as well as a circuit. The idea is that the PCB will form the chassis of the stoRPer, with the motors clamped to the PCB chassis using some 3D-printed parts. Therefore, you need to be capable of placing components and creating PCB geometry accurately in order for everything to fit together. This chapter also looks at how you can check whether PCB and 3D-printed models will fit together before printing or sending out the PCB for fabrication.

One of the first jobs is to create a Pico symbol component in the Symbol Editor. We covered creating symbols in Chapter 3, *Libraries, symbols and footprints*, and while we won't go over all the steps here, you can apply the skills you learned in that chapter to create your Pico component. We decided not to include the Pico's three debug pins on either the schematic symbol or the PCB footprint. This was partly because, across the different Pico models, they are physically in different positions on the board and, as we intend to have a Pico mounted onto this project, we can still interact/wire to the debug pins if needed. As such, we laid out a simple 40-pin component in the Symbol Editor and brought it into the Schematic Editor, as shown in **Figure 7-2**.

After quickly connecting all the ground points, we set about connecting four hierarchical sheets, each with an L9110S motor driver IC-based circuit inside. Chapter 6, *Schematic organisation* covered working with hierarchical sheets, but you can see the circuit layout in **Figure 7-3**. Each of the four motor drivers has its own sheet and has two pins broken out. We've connected these sets of pins to the Pico symbol using labels A1, B1, A2, B2, etc. The rest of the Pico's pins are broken out and connected to some multi-pin connectors, ready to be broken out on the PCB.

For the stoRPer project, we've decided to mount the Pico using the through-hole header pads on the Pico rather than the castellated edge connectors. This means that you won't be mounting the Pico flush to the project, but it does mean that the Pico footprint is thinner. You could also choose to use header sockets to allow the Pico to be temporarily mounted to the PCB.

The header pin pads on the Pico lie in a 2.54mm pitch grid, with the 20 pins on either side being separated by 7*2.54mm. This makes them easy to lay out —

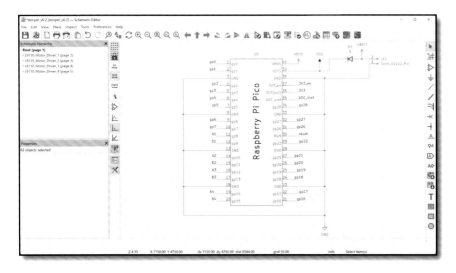

Figure 7-2   A custom Pico symbol with most of the pins broken out

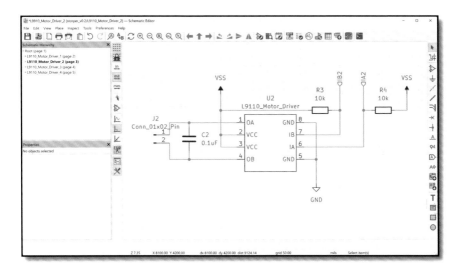

Figure 7-3   The layout of the L9110S motor driver circuit cloned into four hierarchical sheets

simply add pads on a 2.54mm grid in the Footprint Editor (**Figure 7-4**). You'll also want to be able to place a rectangle on the silkscreen layer that accurately shows the position of the board.

Lots of holes

When creating footprints with lots of through-hole pads, KiCad makes it simple: you click to add a pad and then the tool indexes to the next numerical pad for you to place. If you've placed and positioned a lot of pads though, it can be annoying to realise that you need to change an aspect of the pad's properties for all of them. KiCad has you covered, though. As an example, when we made the footprint for a Raspberry Pi Pico, after laying out 40 standard through-hole pads, we decided to increase the internal hole diameter and the overall outer diameter. The Footprint Editor conveniently recognises that this is a common situation and, as such, you can simply change one pad to your desired pad properties and then, with your adjusted single pad highlighted, you can right-click and select **Push Pad Properties to Other Pads...** to make all compatible pads take on the new characteristics.

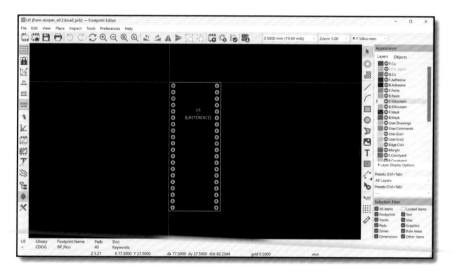

Figure 7-4   Creating the simple yet accurate Pico footprint

Consulting the Pico documentation, you can find a technical drawing and see that the outer edge of the Pico is 51mm × 21mm. You also need to consider the position of this rectangle relative to the pads that you just created. You can see in the technical drawing, for example, that relative to the centre of the upper left-hand pin (pin 1), the upper-left corner of the Pico is 1.37mm higher in the Y axis and 1.61mm over to the left in the X axis. To use this information, go back into the Footprint Editor and place your pointer on the grid point in the centre of the pad you placed for pin 1.

You can press the space bar to set the local origin of the page to be 0,0 at this point. You can check this by looking at the bottom of the screen as you move the pointer, the distance should increase relative to this point. You can then set a user grid to 1mm spacing and use this grid to draw a 51mm × 21mm rectangle. If you then select the rectangle, you can right-click and scroll in the drop-down menu to **Positioning Tools > Position Relative To....** Selecting this, you will see a dialogue box. In the dialogue box, click to select **Use Local Origin** and then adjust the **Offset X** and **Offset Y** by the amounts you derived from the technical drawing (**Figure 7-5**). Note that, by default, the origin corner of the rectangle is the top left-hand corner. Using this method, you can place items with incredible accuracy.

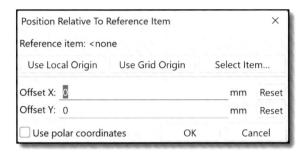

Figure 7-5   Using the 'Position Relative To...' positioning tool to accurately place objects in the Footprint Editor

One thing of note is that despite the stoRPer robot design being relatively simple mechanically — a rectangular PCB — you need to be able to place footprints accurately within the edge cut area. When designing this and other footprints, it's worth considering where your origin point is in the Footprint Editor and placing the device in a known position relative to it. We opted to place the Pico footprint so that the upper left-hand corner of the silkscreen box depicting the edge of the Pico was the origin point on a 1mm grid spacing. This meant that later, when we placed a rectangle in the PCB Editor that represented the edge of the PCB, we could place it in a position such that Pico is dead centre, with the larger box also placed on a 1mm grid coordinate.

After playing with a few test boxes in KiCad, we decided the rectangular chassis dimensions would be 64mm × 86mm. You should use Inkscape to draw a rectangle as you can easily add a 2mm radius to each corner of the rectangle. See **"Carving a path" on page 32** for details on importing graphics. You can easily import a rectangle you draw as an SVG in Inkscape into the edge cuts layer using the **File > Import > Graphics** function.

With the Pico placed and the PCB edge defined, you need to consider the physical mounts for the motors. We suggest using the excellent N20-style geared motors, mounting one for each of the four motor driver circuits. You'll want to 3D-print some brackets to clamp the motors into position, so you need to leave some space for the 3D print material around the motor, and need to take this into account when creating a footprint for the motor mount.

Free Book

The free-to-download book *FreeCAD for Makers* from Raspberry Pi Press looks at the use of the KiCad StepUp workbench. This enables and simplifies importing KiCad projects as 3D objects into FreeCAD, as well as the creation of 3D components for inclusion into KiCad's 3D PCB viewer (**Figure 7-6**).

It's an incredibly powerful suite of tools and is worth exploring. For this project, however, we just wanted to check if the motor clamp we had created in FreeCAD would fit the PCB chassis. You can use **File > Export** and select the **STEP...** option to export a STEP file which can be imported into FreeCAD; however, this will lack the details of the copper layers and silkscreen which you might need to see to check if mechanical components cover aspects of your PCB design. One simple approach that solves this is to export a WRL file. WRL files are file types often used by assets destined for use in virtual reality, but they have the advantage in KiCad that a WRL export contains all the visual details of your PCB.

We used **File > Export > 'VRML...'** to export a WRL file, and then we used **File > Import** in a new document in FreeCAD to import the file. We'd made a simple N20 clamp component which had 2mm radius corners on two corners matching the PCB and N20 motor clamp footprint. While we could have used an Assembly workbench, such as A2plus in FreeCAD, to constrain the clamp in position, for a simple check, we can move the part into alignment to visually check how it looks.

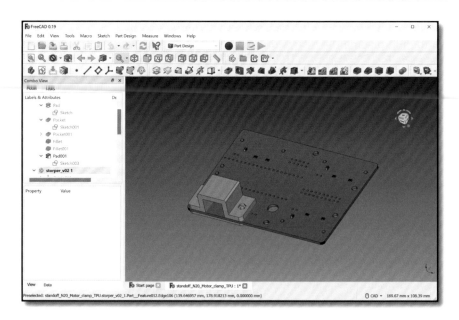

Figure 7-6　The combination of KiCAD and FreeCAD make a great open source toolchain

After some consideration, we created a custom footprint which consisted of two non-plated through-hole (NPTH) mechanical pads placed in-line. These were placed at a distance between centres of 26mm, placed on a 1mm grid spacing. To place an NPTH mechanical hole, you place a regular pad and then press **E** to change the pad type in the **Pad Properties** dialogue. Set each NPTH hole to 2.1mm internal diameter to create clearance for a small M2 bolt.

To finish the footprint for the N20 motor mount clamps, add a silkscreen rectangle set to the dimensions of the base of the 3D-printable mount design (**Figure 7-7**). To add these mounts (which aren't connected to anything electronically) click the **Add a footprint** tool icon and select a footprint in a similar manner to how you would place a symbol in a schematic.

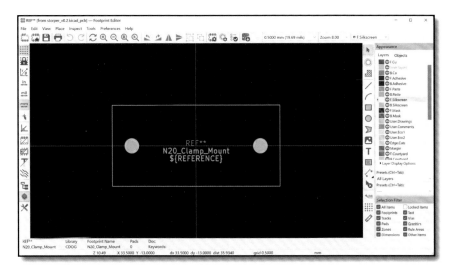

Figure 7-7 The mechanical footprint that will mount the N20 motor and clamp

Quick Tip

As the stoRPer design evolved, we used simple rectangular boxes drawn in KiCad on either the **F.Silkscreen** or the **User.Comments** layer as guides and visual aids.

Adding and removing text-based elements to a silkscreen layer is reasonably straightforward in KiCad. On more technical PCBs (as opposed to artistic PCBs) we often lay out PCB designs with little regard for the silkscreen and then sort the silkscreen layer out later in the development. Often, one of the first tasks is to remove unwanted elements on the silkscreen that have been automatically placed by using default library footprints.

You can select the correct silkscreen layer (often the front silkscreen **F.Silkscreen**), and for items such as footprint reference annotation, click to select them, then move them or press the **Delete** key to remove the item. It's common for this reference to not be placed optimally and may sit under or across other parts and components.

The annotated reference is formed from both the automatic annotation of the schematic during the footprint association process and the type of component it is, so **R** for resistor, **C** for capacitor, **J** for connector, **U** for IC, etc. As they replace the placeholder **Ref*** designator, they are independent of the main footprint design and, as such, can be removed with ease.

If, when tidying the PCB design, you want to move a part of the silkscreen design of a footprint, you will need to edit that in the Footprint Editor. KiCad makes it easy to edit the footprint and apply the changes just to the individual footprint within this project rather than pushing the changes to the global footprint library. With a target footprint selected in your PCB, press **Control** and **E** to open the footprint in the Footprint Editor. You should see the footprint in the editor with a message in the upper left-hand corner of the window that reads **Editing J4 from board**. Saving will update the board only, where **J4** will be the reference of whatever footprint you have opened (**Figure 7-8**). You can now make any changes to the footprint that you require, including deletions or changes to the graphical silkscreen elements.

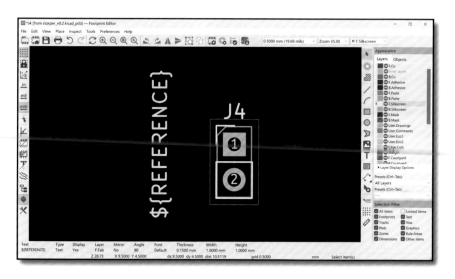

Figure 7-8   Editing a footprint with the component selected and opened in the Footprint Editor from the PCB Editor gives the option of only editing that individual instance of the footprint

You will often want to add text-based components to board designs, and KiCad makes this straightforward. Click the **Add a text item** tool icon and then click in the PCB design. The **Text Properties** dialogue lets you insert text, make changes to the font and size, as well as change the orientation of text. One recent addition to KiCad is the **Knockout** option (**Figure 7-9**). If you input some text into the **Text Properties** dialogue and click the **Knockout** check-box, then the text will be created as a solid silkscreen block with the text removed. It's a great effect, looks smart, and is very readable — a welcome new feature (**Figure 7-10**).

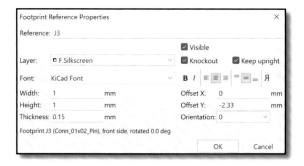

Figure 7-9 The **Text Properties** dialogue where you can set text features, including the new 'Knockout' feature

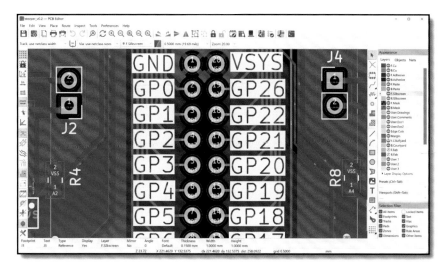

Figure 7-10 A new feature in KiCad 7 is the ability to add knockout text items, where the text is subtracted from a small block on the silkscreen layer

Finally on adding text, sometimes you might like to add text to the silkscreen layer as a graphic rather than directly as text. You previously learned about importing graphics for creating edge cuts geometry and for importing logo graphics from Inkscape. If you use the text creation tools in Inkscape and then directly try to load them as a graphic element, this will fail, as KiCad SVG import doesn't recognise the text elements. This is easy to rectify. For example, you can create your stoRPer text in Inkscape, then, with the text object selected, click **Path > Object to Path** (**Figure 7-11**). Next, edit the document

properties in Inkscape so that the document is the size of the text object —
then save the file as a standard SVG. In the PCB Editor, select **File > Import >
Graphics** to import the file, ensuring to select the correct **F.Silkscreen** as the
graphic layer. The text graphic then imports correctly and can be placed in
the design where required.

Figure 7-11   Converting a text object to a path in Inkscape ready for import into KiCad

You've now got everything in the right place, so you can send the file off to be
manufactured — confident that your robot will fit together.

Chapter 8

Different substrates

A PCB is a copper sandwich — pick the right filling

So far in this book, you have designed your PCBs and had them manufactured using the most common PCB material — FR4. On FR4, copper traces sit on top of fibreglass, and this is what most people think of when they hear the term 'PCB'. However, it's not the only option, and most of the PCB fabrication houses offer a variety of different materials that they can make a PCB from. The circuit is typically still copper, but these traces sit on top of other materials. Let's look at some options.

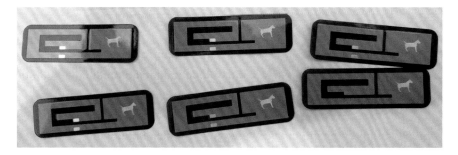

Figure 8-1   Some small polyamide flexible PCB antennas fabricated by OSH Park

FR4 is a standard of material for the dielectric, the non-conducting part of a PCB. FR4 is a type of composite fibreglass material, made up from fine glass reinforcing fibres and epoxy resin, and is very non-conductive. 'FR' is an abbreviation for 'fire-retardant', which is one of the many benefits and safety features of the material. Whilst fibreglass composite materials have inherent fire-resistant qualities, FR4 has bromine added which will reduce the spread of fire.

Gold or silver

There are lots of other choices you can make when getting a PCB fabricated. One choice is the type of surface finish applied to any exposed copper pads. These take the form of differing types of covering or plating, which act to both allow the easy fitting of components, solder, or solder paste, whilst stopping bare copper pads from oxidising if left uncovered.

Hot Air Solder Levelling (HASL) is a common option where the board is dipped fully into molten solder, removed, and excess solder driven off with — you guessed it — hot air. This results in all copper areas being covered in a flat, thin layer of solder. It's an older technology and has been used in PCB fabrication for multiple decades and, as such, it's usually an affordable option. It is easy to solder onto and offers good protection against oxidisation. It has a long shelf life, so if you don't get around to populating your PCBs for a long time, they will still be in good condition. Finally, HASL is often offered in lead or lead-free options.

Electroless Nickel Immersion Gold (ENIG) is a newer technology that applies a two-layer coating to the exposed copper parts of a PCB. The layer directly on top of the copper is a thin layer of hard nickel which creates a barrier stopping the copper from oxidising. To stop the nickel plating from oxidising, a fine layer of gold is immersion-plated over the top. When components are soldered onto ENIG surfaces, they create a strong bond with the nickel layer, and therefore the copper underneath. ENIG is also suitable for covering larger planes on the PCB, so has become a popular choice.

There are other surface finishes, such as *Immersion Tin*, *Organic Coating* (OSP), and *Immersion Silver*, but HASL and ENIG are most common.

For example, JLCPCB has leaded HASL selected as default, but you can switch to lead-free HASL or ENIG. OSH Park PCBs are all created with an ENIG surface finish. Over on PCBWay, you can select from a larger range that includes HASL, ENIG, OSP and more — you can even specify to have no coating whatsoever.

As well as being essentially fireproof, FR4 has a low thermal expansion coefficient — this means that it won't expand or contract very much in hot or cold environments. This is important because if a PCB expands or contracts a large amount, then it is likely that traces and components on the board will be damaged, crack, or disconnect. Most PCB houses will offer a choice of thickness of FR4 boards, allowing you to choose a thickness and strength suitable for your application.

As FR4 is the most common PCB material, there aren't many special considerations or changes needed when designing a PCB for FR4. If you want to set up KiCad to display your target board thickness when using the 3D viewer, you can change this by adjusting the dielectric thickness value via the **Board Setup**. This dialogue is found at **File > Board Setup** in the PCB Editor, then you need to select the **Physical Stackup** option from the list, and then adjust the **Dielectric 1** thickness variable (**Figure 8-2**).

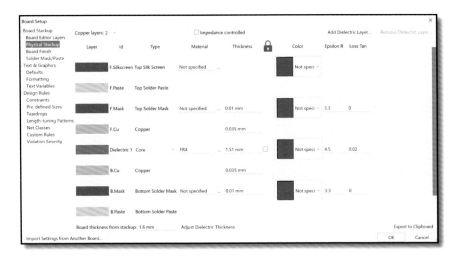

Figure 8-2 Changing the thickness of the 'Dielectric 1' layer to change the overall thickness of the PCB design

Wibbly wobbly

Flexible PCB designs are commonplace in modern electronics for many use cases. Probably the most common are simple flex connectors. These have obvious advantages, in that they can connect components or PCB modules in different locations in a system. Flex connectors can be inserted into specific clamping sockets or directly soldered onto copper pads. They have the benefit of being able to fold and bend around, allowing complex layups of PCBs which do not need larger header pins or sockets.

There are numerous types of flexible PCB substrate, with the most common being polyamide film. Traces and pads work in a very similar manner to any PCB, in that they are a thin layer of copper. Flexible PCBs can become quite complex to design for, and many PCB fabrication houses will have options for custom layer materials in flex PCBs. These can also include rigid sections, which means you must supply a design that can distinguish different substrates within it.

If you plan to use this technology, we'd recommend you read any guidance your PCB fabrication house has and to speak to them directly to explain your concept. It can be as straightforward as laying out a regular PCB design, exporting some Gerbers as you would with a rigid PCB, but there may be specific considerations.

Copper Conundrum

Copper weight describes the thickness of copper at any given point on a PCB. It's often expressed as ounces per square foot, so a 1oz/ft^2 copper weight will be thinner than 2oz/ft^2. Copper weight can be important in terms of designing PCB traces, as the weight affects the depth of the trace. Different trace widths and weights may need to be used when considering the amount of current that certain parts of a circuit may be passing. Similarly, track impedance and RF qualities of tracks may well come into play, as well as track sizes, when trying to match lengths of track in more complex PCB designs.

Most common copper weightings are 1oz or 2oz, and many PCB houses will offer these as a choice. If you require a precise copper weighting, it's possible, at a price, for some PCB fabricators to offer more bespoke weight of copper by plating or etching extra material to or from the board.

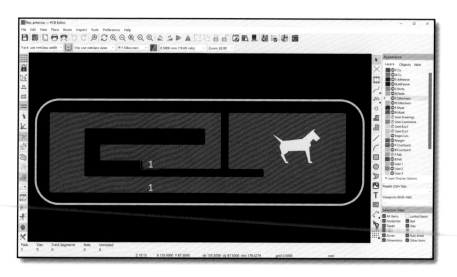

Figure 8-3   The flexible antenna design in KiCad

A good use case for flex PCBs is a flexible antenna. We'd found a paper online with an image and dimensions for a dual-band 2.4GHz and 5GHz patch antenna, which piqued our curiosity, so we set about laying it out in KiCad. We began in Inkscape: we imported the PDF source we had for the design into Inkscape, but it had a gradient fill and therefore wouldn't export correctly as the whole object was full of nodes. If this was a bitmap image, it would have been a good candidate for Inkscape's **Trace Bitmap** function, but as it was a vector file, this wouldn't work. However, as a vector image, it was straightforward to drag in vertical and horizontal guide lines and snap these to the edges of the PDF antenna drawing (**Figure 8-4**).

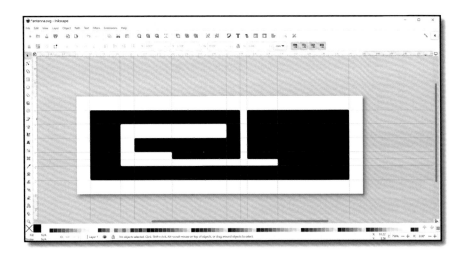

Figure 8-4 Using guide lines in Inkscape to manually trace the antenna design image

Once we had a guide line for every part of the antenna in place, it was a simple job to use the pen tool to draw a continuous line around the part, closing it to form a solid object.

Flexible PCBs can be bent, but over multiple bends, they will start to break. Exactly how often will vary between manufacturing houses, and specific designs.

One thing you can do to prolong the life of the PCB is avoid using sharp 90-degree angles in traces when aiming for a flex PCB as, if the PCB is flexed, the sharp corners can be origin points for tears and failures. Whilst this is not a massive concern for this design, it was easy to add internal and external radial chamfers to the antenna object by selecting the object, then using the **Path effects** dialogue to apply the **Corners path effect chamfer** option (**Figure 8-5**). Finally, we deleted the original imported PDF, resized the document to fit the antenna design, and then saved it as an SVG.

In a new KiCad project, we ignored the usual workflow of creating a schematic and associating parts to schematic symbols and went straight to the PCB Editor. You can directly add components and create traces in the PCB Editor with no schematic in place. For a more complex project, we wouldn't recommend this approach, as you have no means of checking connectivity or creating net connections, but for small simple projects like this, it's easy.

We imported the antenna SVG, making sure to import it onto the Front Copper layer (**F.Cu**). We then used the **Add a Footprint** tool to add two small SMD pads. We positioned these on the points in the antenna design that the origi-

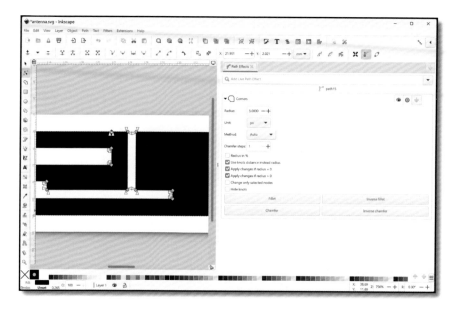

Figure 8-5   Using the **Corners** path effect in Inkscape to add internal and external chamfers

nal design had indicated. With no net connections due to no schematic, these pads will be directly connected to the large copper antenna design that we just placed, but of course, the pads will have no solder mask over them, allowing you to solder on a connecting coaxial cable. All that remains is to then quickly draw another SVG for the outline of the antenna, which we did in Inkscape, but you could just draw in KiCad. With the outline imported to the edge cuts layer, you have a completed design.

To get the design fabricated, we used OSH Park which has a flex PCB offering. One of the nice features of OSH Park is that you don't have to produce Gerber files to upload — you can upload your KiCad PCB file directly to the website for previewing and ordering. Conveniently, project files and Gerber files don't particularly specify the board substrate or thickness, so you don't need to specify a thin flex board design in KiCad. However, you might want to model the board accurately in KiCad, especially if you are using either renders of the KiCad design in promotion or if you are exporting the board 3D model for use in other CAD programs.

You can use the board setup dialogue and the **Physical Stackup** tab in the PCB Editor to emulate a flex circuit. Your PCB fabrication house will have data about all the thicknesses of each layer of their flexible PCB offerings and you can use this to set thicknesses in the board setup. If you just need a close enough PCB view that looks like your flex design, you can simply adapt the major thickness of the board by changing the material to **Polyamide** and setting the thickness of that layer to 0.0102mm (this is the OSH Park flex polyamide layer thickness), as in **Figure 8-6**. You can then set the colours of the top and bottom solder mask layers to transparent by reducing the opacity to zero using the **Custom colour** option. This, in the 3D viewer, will then give you a reasonable approximation of a flex PCB (**Figure 8-7**).

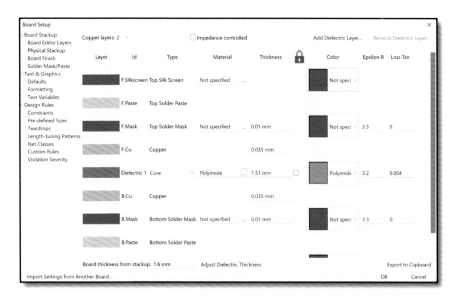

Figure 8-6   By setting the board material, dielectric thickness, and solder mask colours, you can emulate a flex PCB in the KiCad 3D viewer

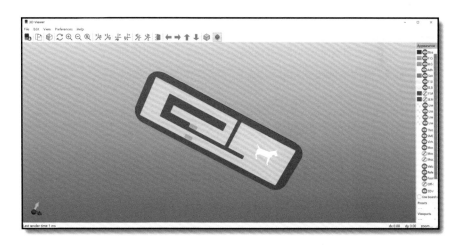

Figure 8-7   A 3D render of a flex circuit

Hard as nails

Often, PCB manufacturers offer metal substrates for PCBs, commonly copper or aluminium. These substrates can be useful when you need to dissipate heat quickly through a system. Often, aluminium substrates can make sense for temperature sensor modules where you want the sensor to be accurate but don't want the board to soak up heat. Another application for metal substrate PCBs is where you want the PCB to act as a heatsink. For example, we had JL-CPCB make and assemble some 1-watt LED modules (**Figure 8-8**).

Running at 1 watt, these LEDs generate a reasonable amount of heat and, to promote their long life, it's useful to have some kind of heatsink. The reverse side of the LED has a large thermal pad which connects to the board, and the reverse of the aluminium board is bare metal. This acts as a heatsink for the LED module and the temperature is reduced. One thing of note is that aluminium and copper substrate PCBs can be bent if they are put under pressure or load. In fact, when your aluminium PCBs arrive in panels, it can be quite hard to remove them without bending the PCB.

Filling holes

Vias, the small plated through-holes that connect different layers of the PCB can be finished in different ways. For many projects, the PCB fabrication house default will be fine, but it's worth looking at the common options offered:

- *Tented vias* are covered with the solder mask, so no solder would stick to them. The hole may or may not be filled, depending on the size of the via. Another benefit of tenting is that you reduce the risk of unintended shorts when boards are being assembled into products or being handled.

- *Un-tented vias* have no covering, so are finished in the selected surface finish in the same way as exposed pads and other copper features. Whilst this may well be fine, there is a risk of accidentally soldering to vias or for short circuits.

- *Plugged vias* are filled in a couple of different ways, or you may have a choice. One way is to fill the via with solder mask; another is to fill the via with epoxy resin. Some manufacturers may only be able to fill vias up to a certain diameter or, indeed, some PCB houses can offer custom approaches where you can ask for all vias of a certain diameter to be filled. The benefit of plugging vias is that the via can't accidentally become filled with solder or other conductive material.

- *Conductive plugged vias* are not the most common choice, but some PCB houses can fill vias with conductive material. This can increase the amount of current the via can pass. There are trade-offs in that the conductive filler may thermally expand at different rates than the other board materials, causing small flexes leading to potential failures. As an example, JLCPCB offers the option to fill vias with conductive copper-filled epoxy.

FR4, flexible, and aluminium are not the only options. With high-speed designs, designs running at microwave frequencies, or very special applications like medical devices — there are other substrates available. Rogers, PTFE, Teflon, Copper Core. These are specialist materials, and we won't look at them here.

Figure 8-8   A small aluminium LED module

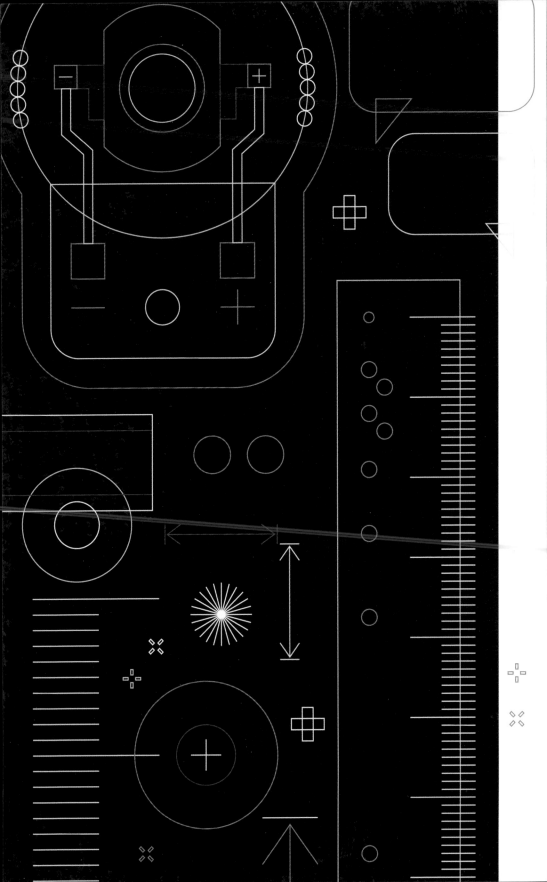

Chapter 9

Finding a PCB manufacturer

Picking the best fabricator for your project

In the last chapter, we looked at the different substrates and other hardware options that are available across a range of PCB fabricators and PCBA services. In this chapter, we are going to generally look at the range of services that are available and look at what some of the companies need from you to make your PCB projects. We'll also mention a few of the quirks we've found in some services that had us scratching our head.

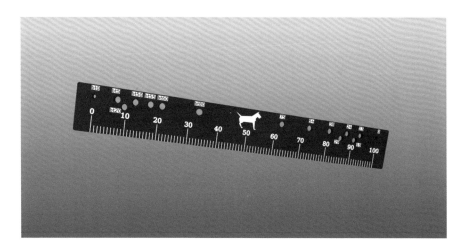

Figure 9-1   The small PCB ruler project provided some interesting challenges to PCB services

There are many well-established companies capable of creating quality PCBs with and without assembly. However, they each have different specifications and tolerances. In fact, that's often a good place to start when comparing or looking for a service to make your project. Questions to ask yourself include:

what minimum clearances and track widths do we need? What is the smallest hole diameter? How accurate do we need things?

Over on OSH Park, they have a great collection of documents relating to the services they provide (**Figure 9-2**). The *Drill Specs* page details the minimum and maximum hole sizes, sizes for annular rings, and the via plating specification. Be aware that all these specifications are different across OSH Park's various services, changing, for example, between the two-, four-, and six-layer options. The minimum track width specification on the OSH Park site is listed as 0.006", with 0.006" clearance on the standard two-layer boards, moving down to 0.005" in the four-layer board offerings (it's common to see limits given in thousandths of an inch because, apparently, you can have both metric and imperial at the same time if you try hard enough).

You can find this and all the other details on the *OSH Park KiCad Design Rules* page at **hsmag.cc/oshparkrules**.

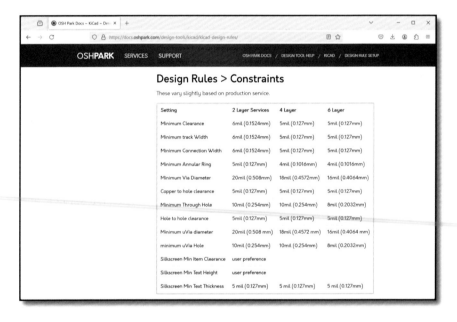

Figure 9-2   Technical details of the OSH Park PCB services

If you have questions about the OSH Park services, they have an excellent track record in communication. You can email the support email address, and they will get back to you offering advice. We have even had the OSH Park team open a KiCad project file, and they have fixed problems and then taken the time to teach us solutions. Many of the services offer online chat portals (**Figure 9-3**). These can be useful when trying to negotiate problems or challenges with a service.

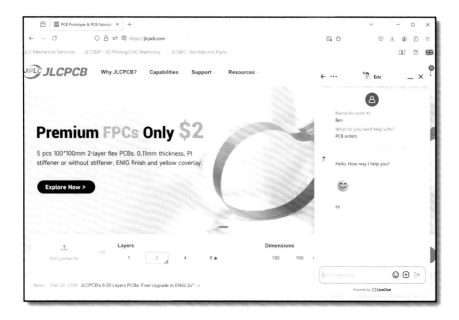

Figure 9-3   JLCPCB's chat portal

Different PCB manufacturers put the information in different places. Another PCB service, DirtyPCBs, bundle their specifications and tolerances information on their *About* page (**Figure 9-4**). They list a 0.006" minimum track width and clearance across both their two- and four-layer boards and cite their other specifications and tolerances. The DirtyPCBs service was designed as a minimal service with its emphasis on cheap, therefore, it has no chat service, and it's difficult to contact the company to ask questions. If you have challenges here, it can be a better option to use search engines and find forum conversations about the service to try and get the answers you need.

We have, of course, used JLCPCB a fair amount in this book. JLCPCB have an exhaustive list of their specifications over on their *Capabilities* page (**Figure 9-5**). They can offer up to a whopping 20 copper layers in their PCBs, with track widths and clearances a default minimum of 0.005" in the two-layer offering, moving to 0.0035" for four-layer options and more.

Similar specifications are available from PCBWay. An advantage of this service is the maximum dimensions of PCBs they can fabricate are 1100 × 500mm, whereas, for example, JLCPCB are 500 × 400mm. One thing of note about PCBWay is that you specify the board and the board dimensions and add it to your shopping cart prior to uploading Gerbers, which can seem slightly counter-intuitive compared to other services.

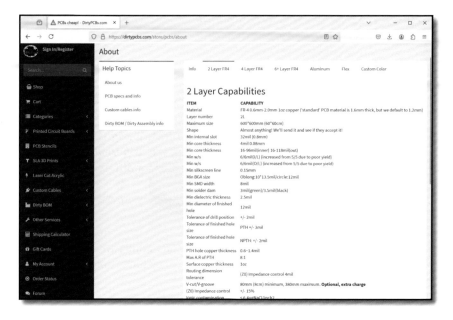

Figure 9-4   The DirtyPCBs PCB specification is in amongst a large About page

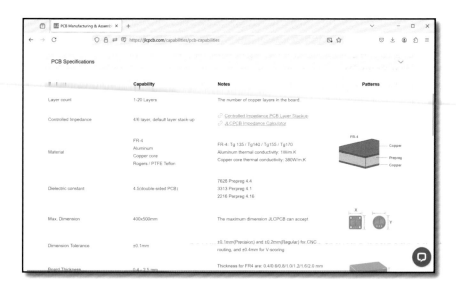

Figure 9-5   JLCPCB's specifications are found on their *Capabilities* page

With all these services, it's worth contacting them using the chat function or emailing if you want to check specifications.

Rules Rule

While developing a PCB project, if you have a particular fabrication service in mind, you can ensure your board design's compatibility by setting up the **Design Rules** for your board to match the service. The **Design Rules** is nested under the **Board Setup** window we used in **Chapter 8,** *Different substrates* to set up the different physical stackup characteristics for different substrates.

In the PCB Editor, navigate to **File > Board Setup** and then open the drop-down menu labelled **Design Rules**. In the first section, **Constraints,** you can set up limits for the minimum clearances, minimum track widths, and other limits relating to the copper regions. You can also adjust the minimum via sizes, hole sizes, and clearances, and the minimum dimensions for text objects on the silkscreen layer.

The next item down in the **Design Rules** drop-down menu is the **Pre-defined Sizes** tab. You used this earlier in the book to set track widths for projects. However, as a reminder, you could set this up at the beginning of a project whilst considering any limitations or constraints your target PCB service has. Also, notice that you can import the settings from a previous project — this is helpful if you set up a project for a particular PCB service specification.

Jumping to the bottom of the **Design Rules** drop-down menu, you can see the **Violation Severity** section. This section sets up how the **Design Rule Checker** (DRC) tool responds if any of the rules set for a project are broken. As a primary word of warning, think very carefully before setting any of these to 'ignore'. It may well be that for a current project you don't mind some of these issues, but it's possible under different circumstances or projects, an ignored error could be critical.

Back in the PCB Editor window, to run the DRC at any point in your project, you can either click the **Show the design rules checker window** tool icon or you can select **Design Rules Checker** from the Inspect drop-down menu. Once the window is open, you can then click the **Run DRC** button for the PCB to be checked against the defined design rules.

Getting an error or a warning doesn't always mean that your PCB project isn't working, but they are simply indications that there is something that hasn't met the design rules. Rules don't always have to be followed, but it's always good to check which rules are broken so you can be sure that any that are broken are broken intentionally. For example, on the ruler PCB design, we got numerous silkscreen errors as the ruler sat over the edge cut geometry, and we also got lots of courtyard errors where the courtyard areas of the mounting hole footprints we had used had overlapped. All of the actual distances between the mounting holes were over the minimum clearance from each other, so neither of these sets of issues mattered — it was worth checking, though.

The DRC, when run, will add small red arrows or markers on your PCB design, highlighting where the issues are located. If you close the DRC window, the markers remain on your design. When selecting a marker when the DRC window is closed, the issue that the marker relates to will be shown in the lower toolbar on the PCB Editor. You can reopen the DRC window and delete single markers or all markers using the relative buttons. Obviously, if you don't make changes to the PCB design and run the DRC again, removed markers will be replaced.

Rulers Rule

Making a PCB ruler is almost a rite of passage in the PCB-making communities. They can be a simple, useful tool, a good business card, or perhaps even perform some extra function. One of your authors (Jo) has an interest in model and high-power rocketry, so he made a ruler which has some accurately placed holes in it, into which you can place a pen. You can then pin the hole at the 0 or 100mm marker and draw circles that match common rocket motor diameters or common Estes rocket body tube diameters. Handy for impromptu cardstock or balsa rocket component-making. It might seem strange to use a PCB manufacturer for this, but it's a very affordable way of getting very accurate 2D designs made.

One slightly tricky aspect of creating a PCB ruler in KiCad is how to draw a graduated line to give the ruler its measuring graphic element. You can use the excellent open source Inkscape to solve this in combination with KiCad's SVG import abilities.

In Inkscape, draw a straight line using the pen tool and use the height and width settings to set it to the length of the ruled section you require. For a compact ruler PCB, you could go with a 100mm length. Next, select the line and then click **Path > Path Effects**. The **Path Effects** dialogue box should open on the right-hand side of the screen — there should be a search bar at the top of this dialogue. Type in 'ruler' and select the **Ruler Path Effect** that appears in the results. This, in turn, should launch the **Ruler Path Effects** dialogue. Set the **Units** to **mm** and then set the **Mark Distance** to 1. This should then add a graduation line one millimetre apart along your line. Next, you can set the length of the major length for the longer graduation lines and the minor length for the shorter lines. Finally, set the **Major Steps** to **10**. You should now have a ruler graphic with a longer marker every 10mm.

You could use KiCad to add the text to mark the numbers on your rule graphic, but it's easy to use Inkscape's align and distribute tools to bring a line of text labels into alignment with the ruler. You can then resize the document using document properties to the size of the design and save it as an SVG to be imported to the silkscreen layer in KiCad in the PCB Editor. Go to **File > Import Graphics**, and then, in the dialogue, navigate to the SVG, set it to import to the correct layer (in this case, the front silkscreen), and make sure the scaling is set at **1**. It's worth setting the PCB Editor grid to **1mm** so that you can align other elements to the ruler design well. Finally, you can be lazy and leave the baseline in the ruler graphic and then used the edge cuts geometry to remove it — this ensures that the silkscreen ruler lines run right to the edge of the PCB. However, you can remove the original line back in Inkscape.

When you have your path effect applied, you can select the entire ruler graphic and then use **Path > Object to Path** to convert the path effect into regular paths. Then, using the node selection tool, you can select the bottom line and delete it. As it's difficult to grab the nodes at the end of the baseline, you can zoom in and then bend the line away by dragging the baseline in between two of the graduation line nodes. Once the line is bent away from the graduations, you can click it and delete it.

Diving deeper

OSH Park have an excellent track record in supporting and promoting open source projects, and they have an iconic purple solder mask finish which is very visible across lots of maker/hardware hacker projects. However, probably one of the main reasons that OSH Park have often featured as a service is that you can directly upload KiCad PCB files to their website — you don't have to go through the process of making compatible Gerber files. This makes the service easy to use. If you are reading this after a new milestone version of KiCad has been released, you might find that it takes a little while for the OSH Park website service to become compatible, but you can also upload a zip file of Gerbers in the same way as other services in the interim. OSH Park's guidance on Gerber set up and requirements is available at **hsmag.cc/Gen_Gerbers**.

Beyond the standard OSH Park purple offering, there are options for the After Dark finish (black substrate and a clear solder mask), a lighter 0.8mm board with a heavier 2oz copper layer, and a flex option. Whilst they offer good quality, they are incredibly affordable when working with smaller PCB designs. Finally on OSH Park, they are excellent at communications. If you need to raise a ticket to ask a question, they go above and beyond.

Different fabrication houses have different needs around the files and file formats that you upload. One area we have noticed creating issues is the DRL or drill files. In KiCad, you can create either a pair of DRL files, one containing the non-plated through-holes and another containing the plated drill holes, or you can merge these two files into one. JLCPCB wants these files supplied as a pair, whereas if you upload Gerbers with two separate DRL files to OSH Park, you get an error message, but it conveniently will merge the two files online and solve the issue for you.

Some fabrication houses want there to be drill files even if the PCB has no drill holes in the design. This caused an interesting issue when we designed the flex PCB antenna example in **Chapter 8, *Different substrates***. We exported the design Gerbers and DRL files even though the design contained no drilled holes — this was just because we wanted to upload the Gerbers to a range of services to see how they rendered and get quotes.

With JLCPCB, when we uploaded the zipped Gerber file for the flex antenna design, the preview would ignore the edge cuts geometry (**Figure 9-6**), so the curved corners of the design would incorrectly disappear, and the board would appear as having square edges. Chatting to the online chat service, they confirmed that they could see the edge cuts layer and assured us that if we placed the order, the board would be cut correctly.

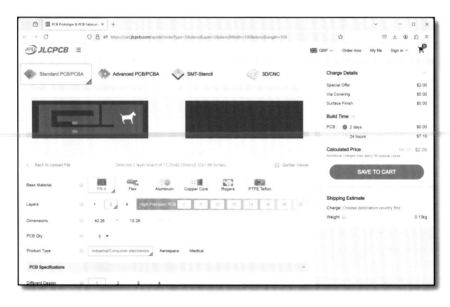

Figure 9-6 Edge cut geometry problems on JLCPCB

We tried playing around with the Gerbers and we also posted the issue on the KiCad forum for discussion. It seemed that others generating their own Gerbers from the project would get a correct render on the JLCPCB site. The difference we spotted was that they weren't including any drill files. Re-uploading without drill files and the correct board outline and edge cut geometry rendered correctly. As part of this process, we discovered that OSH Park didn't have this issue and rendered the board correctly at upload.

The moral of this particular story is that Gerbers aren't standard, so be sure to check what you need, and be prepared to talk to the PCB manufacturer if things don't look right.

We had another issue relating to empty layers when looking at different fabrication services. When uploading the design to PCBWay, although it rendered correctly, it would throw an error with the upload because the Gerber file for the copper layers contained no copper (**Figure 9-7**). Obviously, this is very unusual for a PCB as copper is usually the conductive layer connecting components and more. The ruler project shows that with a more artistic use of PCB fabrication, it's possible to cause headaches for fabrication houses.

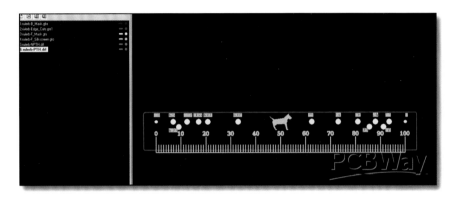

Figure 9-7   The ruler project rendered on PCBWay

We had a problem with a recent project when submitting the aluminium substrate LED module project to JLCPCB services. The 1-watt COB LED we had identified in the JLCPCB parts library had a diagram on the datasheet of the LED which had polarity markings on the two semi-symmetrical flat SMD pin connectors. we'd seen these LEDs in real life — they have an etched - and + in these metal connectors.

When we uploaded to JLCPCB, the website rendered the PCB with the components placed and the JLCPCB 3D model of the LED had no polarity markings. we presumed it would be correct, and as only ordering a small number and the LED is a large part, it wouldn't be too onerous to swap them around if they arrived incorrectly. After ordering, the process was halted, and JLCPCB contacted us to discuss and check the polarity of the LED. We had to point out that it was an issue with their 3D model that meant it was impossible to tell if it was rotated correctly (**Figure 9-8**), and an engineer at their end would only be able to tell when they physically went and looked at the package. In the end, the aluminium LED PCBs were correctly manufactured.

Figure 9-8   The corrected rendering of the LED on JLCPCB

The main takeaway point for working with any PCB service is that most things are achievable with good communication, which increasingly becomes, in combination with the physical specifications, a valuable deciding factor in choosing which service to use.

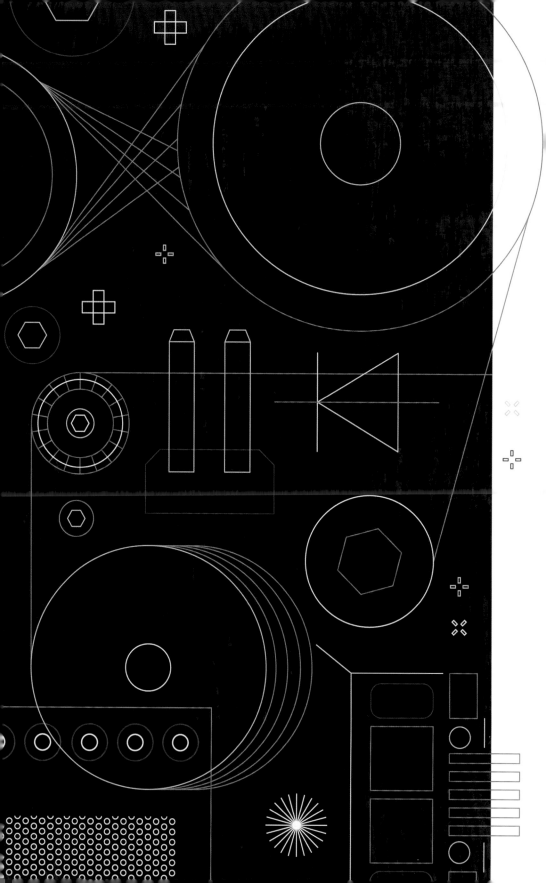

Chapter 10

Making a smart stepper motor

Extend the RP2040 circuit to include a motor driver

Urumbu is a mechanical concept created by Neil Gershenfeld of MIT to simplify the process of creating multi-axis machines (such as 3D printers). The current go-to standard for is to feed Gcode line by line to a controller. This, in turn, drives the individual stepper motors to move the relevant axis. It's a solid system, but it harks from an era where parallel processing capabilities were rare and incredibly expensive.

Figure 10-1   The finished PCB attached to a NEMA 17 stepper motor

Urumbu is interested in streamlining the making of machines, reducing both the cost and the complexity. In simplified terms, it essentially uses stepper motors (or theoretically other actuators) that have been adapted with an embedded microcontroller to run directly via USB. This means that, potentially, you can sidestep using G-code.

Imagine building a machine where you parametrically define the output object, and the script directly calculates the geometry of the form and directly controls the rapid prototyping machine connected to a convenient USB hub. You can see an example of an Urumbu-style controller in **Figure 10-2**. This chapter won't cover all that — it's just going to look at how to build a PCB to control a stepper motor — but you can find out more at **hsmag.cc/urumbu**. You can also look around the Fab Lab depository, where you can find projects that have used the Urumbu approach, like this excellent pointing machine: **hsmag.cc/point**.

Figure 10-2   A NEMA 14 motor with a CNC-milled SAMD11-based board attached

You'll need to create a separate copy of your base KiCad project to work on: first, open the Minimal RP2040 project you created back in Chapter 5, *Designing an RP2040 board*, and then click **File > Save As**. Create another folder on your system and save the project into it with a new name. If you then open this folder in a file management application, you will see that all the KiCad-generated project files (the SCH and PCB etc.) have been renamed to the new project name. You can also tidy and delete files which are specific to the old project and not relevant to the new project. For example, you won't be using the Gerbers, the CSV position, or BOM files, as these will be different for the new project. Similarly, the edge-cut SVG that you created in Inkscape for the Minimal RP2040 project won't be used, so you can delete it. Make sure that you are in the right project folder before you start deleting files!

As the concept for Urumbu stepper motors is to use USB for control, RP2040 is a great candidate for powering a driver board.

Sizing up

There have been examples in the Urumbu community using NEMA 14 motors, which are convenient in the fact that they can often be controlled and powered by USB 2.0 and above. However, most small experimental rapid prototyping machines tend to use the larger NEMA 17 class of stepper motor. Check out any smaller home- or office-use 3D printer, smaller desktop CNC router, or hobby pen plotter and you'll find NEMA 17.

With NEMA 17 as the target, the first port of call is to find some mechanical dimensions and make some fundamental decisions. Looking at datasheets for NEMA 17, you'll find that the outer dimensions of the package are 42 by 42mm, and that they have a set of M3 bolts through the assembly in the corners of a 31mm square. Because you'll use the minimal RP2040 circuit example as the basis of this project, a good starting point is to draw up a NEMA 17 footprint and drop it into the minimal RP2040 design to see how things look (**Figure 10-3**).

You can draw up a quick NEMA 17 footprint in Inkscape and import it to the edge cuts layer in KiCad using **File > Import > Graphics**. After we did that, it was obvious that we would have to reduce the size of some aspects of the minimal RP2040 layout, but not unreasonably so you should be prepared to do the same. You won't need all the GPIO pins broken out, which gives you some easy space-savings. There is just about enough room to also lay a motor driver IC and peripheral components on the board, but you'll be better off using a module for the motor driver section. This makes the board compatible with a range of motor drivers. This means you need to extend the board dimensions in one axis. With this basic feasibility worked out, you can set about editing the schematic to create the new project.

In the Schematic Editor, delete all the GPIO breakout header sections as they won't be needed. You can then use the Symbol Editor to create a custom symbol for the motor driver module (**Figure 10-4**). With the symbol created, connect it to the RP2040 using labels to keep the general schematic sections easier to read.

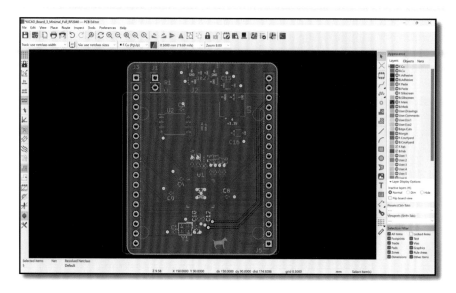

Figure 10-3 A NEMA 17 outline graphic imported into the minimal RP2040 example

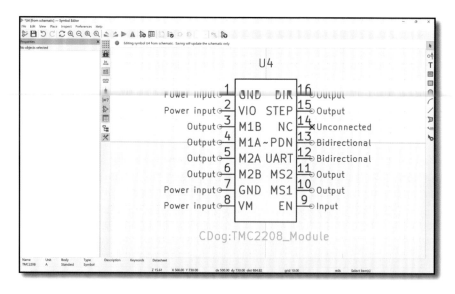

Figure 10-4 A custom symbol for various stepper motor driver modules with a similar form factor

One difference with running a NEMA 17 rather than a NEMA 14 is that although you could, in a slightly limited fashion, run the motor at 5 V from a beefy USB supply, it's likely you might want to run it from a larger external voltage. Most of the common stepper driver modules can do this and have a 'VM' pin into which you can connect an external supply.

To retain the ability to use either USB or VM, you'll need a connector for an external supply and a diode to protect the USB side of the system when the external supply is in use. One of the great things about KiCad is that the workflow of separate schematic symbols (to which you then assign a component footprint) means that you don't have to work out exactly which diode you are going to use right away. You can simply place a diode symbol, wire it into the schematic, and consider the package later (**Figure 10-5**).

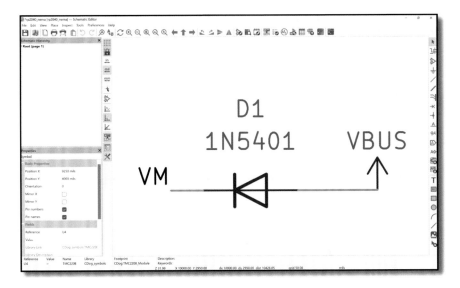

Figure 10-5   KiCad's workflow often lets you add generic component symbols and worry later about which actual component to use

Breaking out

In addition to the motor driver, you'll want two header sockets connected to GPIO and ground in case you want to attach switches to act as limit switches; this gives you feedback and control options for any machines that you might develop with these motors. Again, you can simply add these to the schematic.

With the adapted schematic largely complete, you can set about making decisions on component choices and check which previously used components are available. This is where things can get very tricky and time-consuming when using PCBA services. At the time of this writing, we were glad to see that the RP2040, the Winbond flash chip, and the 12MHz crystal were still in stock with JLCPCB. We also took the time to check that the smaller components — the capacitors and resistors — were all also in stock.

At the time, the USB socket and the 3V3 voltage regulator we had used previously were no longer in stock. Both are high-turnover items and, at one point in this process, we couldn't identify any 3V3 voltage regulators in any package that were suitable for this project. We also found some challenges in that there would be a regulator listed, but not enough information available either in the listing or the component's datasheet to actually make a decision on whether to include the part. With items like voltage regulators, LCSC (the company that is the back end of the JLCPCB component warehouse) is continually changing and adding items and stock. What can appear a huge problem one day in your search results can suddenly have half a dozen more options in a couple of days' time.

Starting a warehouse

One way to avoid component shortages is to pre-order components to be held ready for use in your project. This is sometimes referred to as a *Virtual Warehouse*, where you can buy an inventory of component stock and hold them until you are ready to place them onto a PCB assembly. This functionality is already built into your JLCPCB account, and once signed in to your account, you can move to the **Parts Manager** page. On this page, you can use the **My Parts Lib** to view and to add to your personal parts library. You can buy both Basic and Extended parts, and you can also pre-order out-of-stock extended parts for when they are hopefully restocked.

For Basic parts from the JLCPCB parts warehouse, you have a minimum order requirement. However, Basic parts are much less likely to go out of stock and, if they do, they are likely to have an alternate similar part available. One thing you need to know, though, is that these pre-purchased parts are only for use in assembly services — you can't suddenly have your library of parts mailed to you as a component order.

If you are creating a project and you think you are going to have a long development time where component stock might be an issue, this can be a great option for your peace of mind.

We eventually found replacements for all non-stock components. The replacement USB socket required a different footprint, but we could download the footprint from EasyEDA and import it into KiCad (see **"Planning ahead: components and footprints" on page 51**). **Figure 10-6** shows the new USB connector we identified; we created the footprint by converting the EasyEDA example on the product page.

The new USB connector had some through-hole chassis components but was listed as an SMD. This concerned me, as the PCB assembly service charges

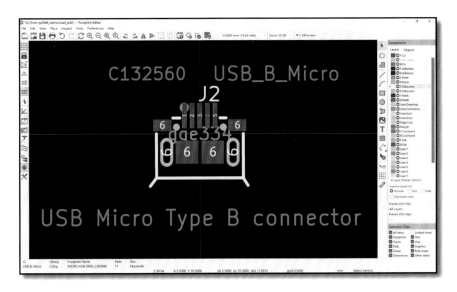

Figure 10-6 The new USB connector footprint

quite a bit more for through-hole than surface-mount, but the part was attached via the single side surface-mount services without problems.

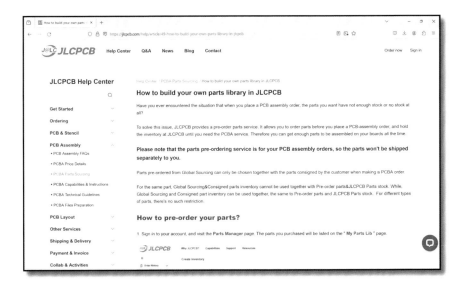

Figure 10-7 When trying to minimise redesigns and faced with component changes, you'll spend a lot of time using the search options!

Having struggled to identify a suitable voltage regulator in any package at one point, we left the project for a couple of days. Later, we found different stock available in the JLCPCB parts library and managed to find a drop-in replacement regulator which would sit on the same SOT23 footprint. It's definitely worth triple-checking the footprint and pinout of any swapped components to ensure that your wiring still works.

With most of the problems solved with regards to components, you can set about editing the PCB layout. You'll need to make the minimal RP2040 layout more compact to fit within the NEMA 17 footprint, and so you should move the actual RP2040 and crystal upwards, decreasing the distance between it and the USB socket. Sometimes, in reworking a PCB like this, the grab function is quite handy, where you can select a track, or a selection of tracks and components, and then use the **G** hot key rather than **M** and, instead of simply moving the objects, they are grabbed and the track connectivity remains which the tracks can move. This rarely results in a neat set of tracks in our experience, but it can be useful to create a quick new routing which you can then manually edit.

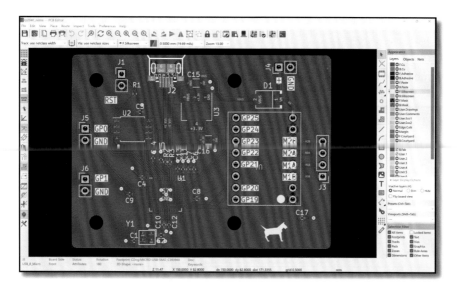

Figure 10-8   The complete Urumbu RP2040 PCB layout

With the PCB design complete, the next step is to create the Gerbers, BOM, and positional files for the project and upload it to JLCPCB. After a short production and delivery of the assembled PCBs, we created a standoff design for 3D printing using FreeCAD, and then the boards simply attach to a NEMA 17 using some longer M3 bolts. If you are interested in replicating these boards or playing with RP2040 Urumbu-style approaches, download this project from the book repository **hsmag.cc/kicad_book_files**.

Whoops-a-daisy

In the spirit of failing out loud, I'd like to share a mistake I made in the production of this PCB. Using motor driver modules rather than designing around a particular motor driver IC meant I had lots of flexibility and redundancy even if certain motor drivers were out of stock. The motor driver boards like the TMC2208 module, the DRV8833 module, and the A4988 module all share a common footprint with 16 pins, in two rows of eight pins in 2.54mm spacing.

I didn't know the spacing between the rows, but a friend had a module on their desk and I messaged them for some dimensions. They sent me a collection of pictures with callipers held to the board, and the module placed into a breadboard. I quickly counted across the breadboard to see how many columns wide the module was. It's six columns wide across the pin rows.

When laying out the simple footprint for the module, I set the grid to 2.54mm in the footprint editor and then drew one column of eight pins. I then counted across six rows and laid out the second column. Of course, that is an error: counting six rows across makes a module that would span seven columns of a breadboard and is therefore 2.54mm too wide. The simplest things are often the worst!

The challenge with this sort of error is that they are not the kind of errors that can be detected by the DRC system, as connectivity to this simple footprint looks correct to the system. Only when the assembled PCBs arrived did I realise the error. For the small number of boards I had manufactured, the horrid workaround is to slightly angle in the header sockets to bring them close to correct and then insert the module. Crude, but allows me to use the boards. Everyone will make mistakes; if and when it happens to you, don't beat yourself up too much.

— Jo

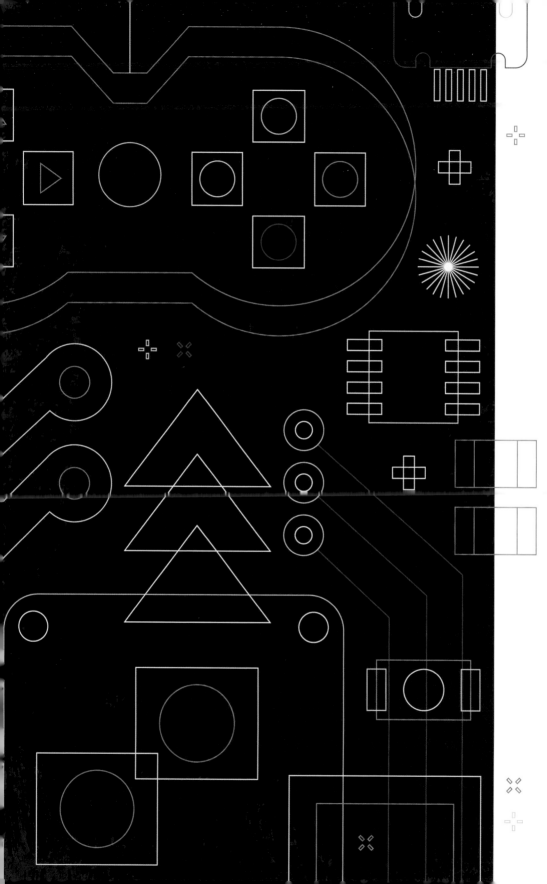

Chapter 11

Making an RP2040 game controller

Let's explore adapting the RP2040 layout to make a USB game controller

Earlier chapters established a reasonable working RP2040 layout, so now it's trivial to create new RP2040 devices. The last chapter used it to make an Urumbu-style motor driver board, and in this chapter, you'll see how you can create a simple USB game controller (**Figure 11-1**).

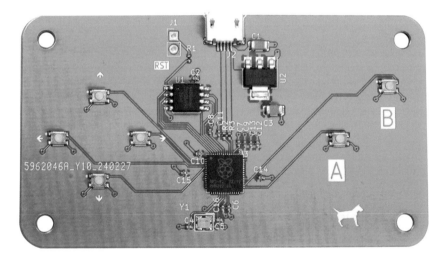

Figure 11-1   The completed game controller PCB

It's largely the same process you undertook for the Urumbu project. Start by making a copy of the Urumbu project and clean out any files in the new project copy that you don't need: the board edge geometry, the Gerbers, and CSV files can all be deleted, as you will replace them with ones generated for the new project. You can also delete all the Urumbu parts on the schematic that you don't need. You don't really need to delete items in the PCB Editor, as when you eventually pull in the updated netlist and bill of materials, you can automatically delete unreferenced footprints, and the new footprints will be brought in.

You'll need to add six tactile buttons to the RP2040: four in a D-pad arrangement and two as A- and B-style buttons. You will want these buttons to be momentary press buttons (also known as *push to make*). You'll connect one side of each button to a GPIO and the other side to ground.

Scouring the JLCPCB parts library, we came across the C221902 button. This part looked a nice size, so we looked at the EasyEDA schematic and footprint. It has four pins and, reading the schematic, we could see that if we connected pin 2 to a GPIO and then connected all the other pins to ground, it would work as needed. Additionally, with the four SMD pads, it should be mechanically strong.

High performance

This chapter looks at creating a gamepad that's easy to understand and extend. However, if you're looking to build a high-performance gamepad, then there are lots of things that you need to consider. Part placement is obviously a large part of it, as you need to be able to press buttons consistently and accurately.

However, another part is the software. The CircuitPython code could be improved, but ultimately, if you're looking for high performance, CircuitPython isn't the right choice. Fortunately, there is another option.

GP2040-CE is a firmware for RP2040-based devices. You can configure it with details of what hardware is connected where. It understands more than just buttons, so you can add analogue inputs as well.

There's documentation on the project website: **hsmag.cc/GP2040-CE**.

With the choice of parts made, we downloaded the footprint from EasyEDA and imported it into KiCad (see **"Planning ahead: components and footprints" on page 51**), then add it to a custom library.

To add the buttons to the schematic, you can create a custom 4-pin schematic symbol and insert it into a hierarchical sheet. Next, wire the GPIO pin and the other pins to ground and then bring out the GPIO hierarchical pin. Then, copy the hierarchical sheet to create six versions, one for each button, adjusting the label and the sheet name as you add each (**Figure 11-2**).

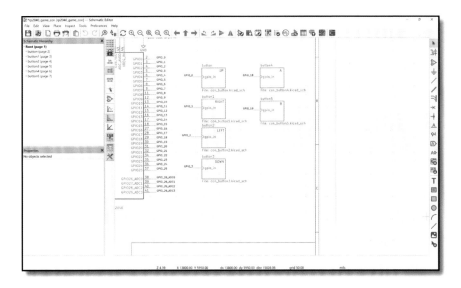

Figure 11-2   Using hierarchical sheets makes it easy to add multiple similar connected schematic blocks, such as the buttons

Next, assign the new footprints to the schematic symbols and began to edit the PCB layout. You should create and import a new board edge geometry SVG in Inkscape with some mount holes (**"Carving a path" on page 32**) before carrying out the usual exporting of Gerbers, BOM, and positional files for JLCPCB services (**Figure 11-3**).

After ordering the boards, one final fun activity on the hardware side of this build is to export a STEP file from KiCad to model around in FreeCAD. To export a basic STEP file from the KiCad PCB Editor, select **File > Export** and then choose STEP as the output format. Note that you haven't added custom 3D models for all the custom components, so obviously the STEP file isn't completely correct, but it serves as a good enough guide to model around in FreeCAD.

In the free-to-download book FreeCAD For Makers (**hsmag.cc/freecadbook**), we explored using FreeCAD and the KiCad StepUp workbench that allows you to create and position custom 3D component models for use in KiCad. we also explored all the skills needed to create all kinds of models. With the knowledge you gain from this book, you could certainly make a controller enclosure like the one we quickly modelled (**Figure 11-4**).

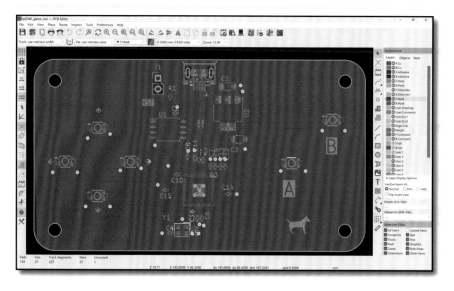

Figure 11-3 The completed PCB layout

Other gamepads

This example should get you started in the world of game controllers, and there are loads that you can look at for inspiration:

- The Arduino Esplora is now retired, but was one of the first hackable game controllers on the market: **hsmag.cc/ArduinoEsplora**

- There's an online community at PCBWay's shared projects site that includes many game controllers, including: **hsmag.cc/PicoGamepad**

- Gamepads come in many shapes. They're usually designed around ergonomics, but you can get a little creative. For example, this maker has built a bat-shaped controller: **hsmag.cc/BatController**

The softer side

Now that you have created the board, it's time to write some code for it. You could write the code in C using the Pico SDK. You could also use the Pico build of MicroPython or CircuitPython. However, since you've created a new board, it's helpful to create a firmware tailored specifically for it — a custom build of CircuitPython. This allows you to do a couple of things. Firstly, it lets you name the specific pins, so rather than using, say, GPIO0, you can use BTN_A. Secondly, it lets you select which modules we want to include. In the case of

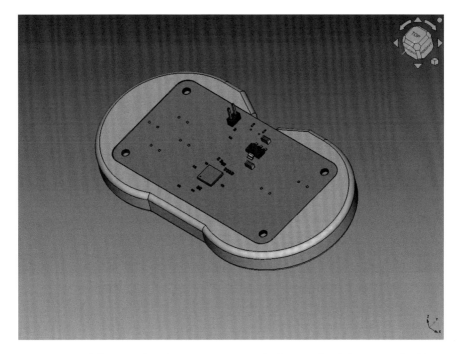

Figure 11-4 Modelling a simple enclosure in FreeCAD to make the controller a little more comfortable to hold

this project, you'll add Adafruit HID, which enables you to use the game controller as an input device.

We found it easiest to build CircuitPython using Windows Subsystem for Linux (**Figure 11-5**), but the general process for creating a build of CircuitPython is given in the documentation at **hsmag.cc/BuildCP**. We won't go through it in detail, so follow that guide to set up your environment.

Lead-free

It's often cheaper to get boards made using leaded solder. However, this might be a false economy. Leaded solder is harmful to both your health and the health of our planet. In the case of a games controller — something that you're going to hold in your hand time and again — it's more important than usual to opt for lead-free solder. Even if only a tiny amount gets on your hands each time you use it, that will still add up over the course of the controller's life and could have negative effects on your health.

Once you have everything set up, you need to create this board. In the directory **circuitpython/ports/raspberrypi/boards**, copy the **raspberry_pi_pico**

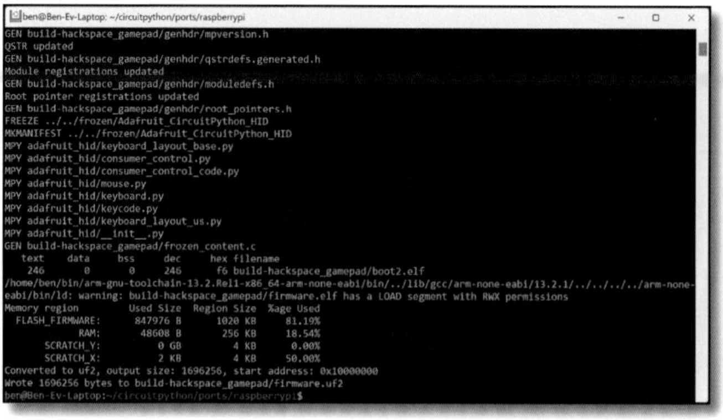

Figure 11-5 Building CircuitPython using Windows Subsystem for Linux

directory into a new one named appropriately for the gamepad. We called ours **hackspace_gamepad**.

There are two files that you need to adjust to customise CircuitPython for your board. Firstly, there's **pins.c**, which you should edit to read as follows, which adds items to the boards module (one item for each button):

```
#include "shared-bindings/board/__init__.h"
STATIC const mp_rom_map_elem_t board_module_globals_table[] = {
    CIRCUITPYTHON_BOARD_DICT_STANDARD_ITEMS
    { MP_ROM_QSTR(MP_QSTR_UP), MP_ROM_PTR(&pin_GPIO0) },
    { MP_ROM_QSTR(MP_QSTR_RIGHT), MP_ROM_PTR(&pin_GPIO1) },
    { MP_ROM_QSTR(MP_QSTR_LEFT), MP_ROM_PTR(&pin_GPIO2) },
    { MP_ROM_QSTR(MP_QSTR_DOWN), MP_ROM_PTR(&pin_GPIO3) },
    { MP_ROM_QSTR(MP_QSTR_BTN_A), MP_ROM_PTR(&pin_GPIO18) },
    { MP_ROM_QSTR(MP_QSTR_BTN_B), MP_ROM_PTR(&pin_GPIO19) }
};
MP_DEFINE_CONST_DICT(board_module_globals,
                     board_module_globals_table);
```

Next, edit **mpconfigboard.mk** to be the following:

```
USB_VID = 0x1209
USB_PID = 0xB182
USB_PRODUCT = "HackSpace gamepad"
USB_MANUFACTURER = "HackSpace magazine"
CHIP_VARIANT = RP2040
CHIP_FAMILY = rp2
EXTERNAL_FLASH_DEVICES = "W25Q128JVxQ"
```

```
CIRCUITPY__EVE = 1
FROZEN_MPY_DIRS += $(TOP)/frozen/Adafruit_CircuitPython_HID
```

In **mpconfigboard.mk**, you define the type of flash chip you have and add any 'frozen' modules you want. Frozen modules can be anything that you want to be included on the build by default (other than the core modules that are automatically included). Frozen modules must be in the **circuitpython/ frozen** directory, but you should find that the **Adafruit_CircuitPython_HID** module is already there.

You can now create your build by going to **circuitpython/ports/raspberrypi** and running:

```
make BOARD=hackspace_gamepad
```

This will compile your code, and you should end up with a **build-hack-space_gamepad** directory. In there, you'll find a **firmware.uf2** file that you can load onto your gamepad just as you would any other UF2 file.

This isn't complete firmware — it's only the programming language. You now need to write a program to get everything working. Fortunately, you have all the modules we need baked in, so there's no need to install any additional modules. Here's some firmware that draws inspiration from the CircuitPython example code. Save it as **code.py** into the root of the device's filesystem:

```
import time
import board
import digitalio
import usb_hid
from adafruit_hid.keyboard import Keyboard
from adafruit_hid.keyboard_layout_us import KeyboardLayoutUS
from adafruit_hid.keycode import Keycode

# A simple neat keyboard demo in CircuitPython

# The pins we'll use, each will have an internal pullup
keypress_pins = [board.UP, board.DOWN, board.LEFT,
                 board.RIGHT, board.BTN_A, board.BTN_B]
# Our array of key objects
key_pin_array = []
# Keycode sent for each button, will be paired with a control key
keys_pressed = [Keycode.UP_ARROW, Keycode.DOWN_ARROW,
                Keycode.LEFT_ARROW, Keycode.RIGHT_ARROW,
                Keycode.A, Keycode.B]
```

```
# Sleep to avoid a race condition on some systems
time.sleep(1)

keyboard = Keyboard(usb_hid.devices)
keyboard_layout = KeyboardLayoutUS(keyboard)

# Make all pin objects inputs with pullups
for pin in keypress_pins:
    key_pin = digitalio.DigitalInOut(pin)
    key_pin.direction = digitalio.Direction.INPUT
    key_pin.pull = digitalio.Pull.UP
    key_pin_array.append(key_pin)

print("Waiting for key pin...")

while True:
    # Check each pin
    for key_pin in key_pin_array:
        i = key_pin_array.index(key_pin)
        key = keys_pressed[i]
        if not key_pin.value:   # Is it grounded?
            print("Pin #%d is grounded." % i)
            # "Type" the Keycode or string
            keyboard.press(key)   # "Press"...
        else:
            keyboard.release(key)
    time.sleep(0.01)
```

As you can see, you can use **board.UP**, **board.DOWN**, **board.LEFT**, **board.RIGHT**, **board.BTN_A**, and **board.BTN_B** in your code. This has a couple of advantages. Firstly, it is more intuitive for programmers. Secondly, if you created another version of the board with the buttons on different pins, the same code could still run on both.

This code is a bit lazy. For example, there's no *debouncing* on the buttons. In practice, we've found that this doesn't cause many problems, especially with the **time.sleep(0.01)** in there. This means it's not the most responsive controller, so if you're playing games where hundredths of a second matter, you probably want to use something different, such as tuned debouncing written in C. However, this controller isn't suitable for that type of game anyway. This is also fairly cavalier with the number of *reports* it sends (a report being a status update sent from keyboard to computer). This will send six of them every loop, which means several hundred a second. Again, this isn't great for performance. However, it works reliably and is easy to understand.

```
Mu 1.0.3 - untitled *                                                    —  □  ×

untitled * »
1 import time
2 import board
3 import digitalio
4 import usb_hid
5 from adafruit_hid.keyboard import Keyboard
6 from adafruit_hid.keyboard_layout_us import KeyboardLayoutUS
7 from adafruit_hid.keycode import Keycode
8
9 # The pins we'll use, each will have an internal pullup
10 keypress_pins = [board.UP, board.DOWN, board.LEFT, board.RIGHT, board.BTN_A, board.BTN_B]
11 # Our array of key objects
12 key_pin_array = []
13 # The Keycode sent for each button, will be paired with a control key
14 keys_pressed = [Keycode.UP_ARROW, Keycode.DOWN_ARROW, Keycode.LEFT_ARROW, Keycode.RIGHT_ARROW, Keyc
15
16 # The keyboard object!
17
18 time.sleep(1)  # Sleep for a bit to avoid a race condition on some systems
19
20 keyboard = Keyboard(usb_hid.devices)
21 keyboard_layout = KeyboardLayoutUS(keyboard)
22
23 # Make all pin objects inputs with pullups
24 for pin in keypress_pins:
25     key_pin = digitalio.DigitalInOut(pin)
26     key_pin.direction = digitalio.Direction.INPUT
27     key_pin.pull = digitalio.Pull.UP
                                                                        Adafruit ©
```

Figure 11-6 The custom build of CircuitPython has everything we need, including pin names and modules

With this code loaded, you should be able to plug the controller into any computer and it will recognise it as a USB keyboard. Press one of the buttons and the computer should recognise that button press just as it would from any keyboard. With this, you can control any game that takes input from a computer.

Creating a custom version of CircuitPython isn't essential when you build a new board; however, once you've been through the process once, it's easy, and makes life a little bit nicer, especially if you're distributing the board to other people.

This concludes our tour of KiCAD with RP2040. PCB design is a fascinating and huge subject, but you should now know enough to design Raspberry Pi Pico-compatible boards, and get them made by a PCB assembly company. You should understand the minimal design and be able to tweak it to your needs. You can add more hardware, or strip it back to its smallest size. You can make the PCB a specific shape to fit in your project, or even make the PCB an integral part of the structure.